摩登样板间 III
新简欧

MODERN SHOW FLAT III
NEW JANE EUROPEAN

ID Book 图书工作室 编

华中科技大学出版社
http://www.hustp.com
中国·武汉

浅谈样板间设计

样板间是地产开发商为吸引目标客户而精心打造出来的理想家居空间。设计师的设计重点在于营造引人入胜的视觉效果、注重表现楼盘的特质及展现目标客户理想的生活品位。

在确定设计主题的过程中，我们首先会与房产商沟通，了解楼盘的特质、当地的文化背景以及目标客户群的需求。关注案子本身的售价，以及开发商想要销售的群体。挖掘这一类目标客户群的使用要求以及兴趣爱好，并思考可能会打动他们的生活场景等。

设计流程的先后次序分别是建筑外观、内部结构、空间设计，前两部分由建筑师完成，后一部分则由室内设计师完成。但建筑师并非就具备空间设计的经验，这会导致空间结构布局的不合理，因此在设计之初我们也会介入建筑的设计，使建筑外观与空间结构布局都更加合理。使得我们在之后的室内设计中更加得心应手。

相比而言我们更注重平面，认为平面比立面要重要，因为我们在做平面的时候就把立面考虑到了，这样当我们把立面拉起来后就已经是很好的作品了。我们常使用"灰空间"的手法。利用室内与其外部环境之间的过渡空间，来达到室内、外融和的目的。用"灰空间"来增加空间的层次，协调不同功能的建筑单体，使其完美统一。改变空间的比例，弥补建筑户型设计的不足，丰富室内空间。

总的来看，我们对于样板间的设计有三个原则。扬长，充分展示自己的优点。避短，通过设计的手法来弥补户型的缺憾，房子或多或少都存在某些缺憾，需要通过设计给予弥补或掩饰。主题风格，样板间设计必须有明显的主题思路或风格，让人们记忆深刻。

台湾大易国际设计事业有限公司•邱春瑞设计师事务所

邱春瑞

摩登时代的Modern家居

我对风格的划分一向持保留态度，尤其在室内设计上。倘若将新中式、新简欧及新现代、新田园放在一起，冠以"Modern"一词结集出版，我亦不觉有丝毫不妥。因为它们都有一个"新"字，只要不钻牛角尖，但凡是新的，理所当然都可以认为是Modern的。

除了"新"，Modern还应该具有以下的特征：新奇的、时尚的、合时宜的。当年我们看卓别林的《摩登时代》，尽管有一些冷幽默的成分，然而，它毕竟阐述了那个时代的精神语言，而我们在设计概念上注入摩登的成分，目的也是显而易见的，你说我们流于俗套也好，你说我们标新立异也好，最重要的是，我们所结集出版的这一系列作品，毋庸置疑，是这个时代，不，是当下乃至再稍后一段时间国内设计的代言。其中的作品，无论新中式的典雅，新简欧的浪漫，还是新现代的飞扬，新田园的清新，都不约而同地在应和着当今各类业主的需求，也展示着一批会思考的优秀设计师对Modern的理解与把握。

中国古语云"识实务者为俊杰"，一个数年致力于设计领域书籍出版的团队，是以极为务实的态度去甄选这套Modern家居系列作品，无论从设计思维还是操作的层面，这套书籍都有可圈可点，可以借鉴、学习之处。

作为设计师，我们通常以上下五千年、纵横千万里、信马由缰、神思飞扬而自诩，而更多时候，也许需要执案自问：我们的思维是否真正Modern了？

香港方黄建筑师事务所

方峻

目录

006-084

- 006 中洲中央公园11-B02样板间
- 012 中洲中央公园1-B02实地样板间
- 018 中洲中央公园二期12-A01户型样板间
- 026 香港信和·中央广场示范单位1
- 030 香港信和·中央广场示范单位2
- 038 上海徐汇园二号楼
- 044 上海徐汇园三号楼
- 048 未来之光样板间
- 052 君悦黄金海岸样板示范单位
- 060 常州乐颐大厦某宅
- 066 长江路九号 No.9
- 072 黑白森林
- 078 苏宁睿城某住宅
- 084 中央原著某住宅

090-156

- 世贸中心邬宅 090
- 锦华园山景水岸样板间 096
- 大奢至尚 102
- 百悦天鹅湖样板间 110
- 常州莱蒙城别墅 116
- 城市时代T1样板间 122
- 简·绎 126
- 桂丹颐景园高层样板间 132
- 宏侨凯旋名门一期15#样板间 138
- 欧式风情遇上现代简约 144
- 恒信·中央公园9C户型样板间 150
- 凯旋枫丹某住宅 156

164-236

- 164 依云溪谷169栋
- 170 君临南山某住宅
- 176 钟鼎山庄
- 180 品味生活
- 186 半岛之恋
- 192 时尚新古典
- 198 中电颐和家园
- 202 中南世纪城
- 208 金丰复式
- 214 时尚圆舞曲
- 218 御江金城
- 224 四季花园
- 230 风华年代
- 236 富丽湾何公馆

244-310

- 怡湖新贵 244
- 低调新简欧 248
- 江景大厦某住宅 256
- 橘郡阳光 262
- 宾王广场样板间 268
- 来宾海德堡样板间 274
- 湖与墅别墅样板间 280
- 中南世纪城9幢 286
- 天正桃园 292
- 嘉宝田花园 298
- 梅溪湖D6区样板间 304
- 公园大地某住宅 310

CONTENTS

006-084

- 006 Central Park, 11-B02 Show Flat
- 012 Central Park 1-B02 Show Flat
- 018 Central Park Phase 2, 12-A01 House Type Show Flat
- 026 Hong Kong Sino• Central Plaza Show Flat 1
- 030 Hong Kong Sino• Central Plaza Show Flat 2
- 038 Shanghai Xuhui Garden, Building No. 2
- 044 Shanghai Xuhui Garden, Building No. 3
- 048 Future Light, Show Flat
- 052 Junyue Grand Coast Show Flat
- 060 One Residence of Changzhou Leyi Building
- 066 Changjiang Road No. 9
- 072 Black and White Forest
- 078 One Residence of Suning Ruicheng
- 084 One Residence of Central Villa

090-170

- 090 World Trade Center Wu's Mansion
- 096 Jinhua Garden Mountain View, Show Flat
- 102 Grand Luxury
- 110 Swan Lake Property, Show Flat
- 116 Changzhou Laimeng Town Villa
- 122 City Time, T1 Show Flat
- 126 Conciseness•Presentation
- 132 Guidan Yijingyuan High-rise, Show Flat
- 138 Hongqiao Triumphal Arch, Phase 1, 15# Show Flat
- 144 When European Charms Meet with Modern Conciseness
- 150 Hengxin•Central Park 9C House Type, Show Flat
- 156 One Residence of Kaixuan Fengdan
- 164 Evian Valley, Building No. 169
- 170 One Residence of La France Romanlique

176-248

- 176 Zhongding Villa
- 180 To Taste the Life
- 186 Love of Peninsula
- 192 Fashion New Classical Style
- 198 Zhongdian Yihe Home
- 202 Central Living District
- 208 Jinfeng Duplex Mansion
- 214 Fashion Waltz
- 218 Riveria Royale
- 224 Four Season Garden
- 230 The Time of Elegance
- 236 Fuli Bay He's Mansion
- 244 Yihu New Aristocracy
- 248 Low Profile New Concise European Style

256-310

- 256 One Residence of Riverview Building
- 262 Orange County Sunshine
- 268 Binwang Plaza, Show Flat
- 274 Laibin Heidelberg Show Flat
- 280 Villa Life by the Lake Show Flat
- 286 Central Living District Building No. 9
- 292 Tianzheng Taoyuan
- 298 Jiabaotian Garden
- 304 Meixi Lake D6 District, Show Flat
- 310 One Residence of Garden Land

中洲中央公园 11-B02 样板间
Central Park, 11-B02 Show Flat

设计公司：KSL 设计事务所
主设计师：林冠成
项目地点：广东深圳
项目面积：153 m²
使用材料：黑檀木、加拿大木纹、黑钢、玉石、夹丝玻璃、壁纸、皮板

Design Company: KSL DESIGN(HK) LTD.
Designer: Andy Lam
Project Location: Shenzhen of Guangdong Province
Project Area: 153 m²
Major Materials: Black Ebony, Canadian Wood Grain, Black Steel, Jade, Wired Glass, Wallpaper, Leather Board

后现代主义风格的室内设计擅长将众多具有隐喻性的视觉符号融入作品中，带给人耳目一新、魅力持久的高品位体验。本案将欧式贵族风格的奢华元素与后现代主义的抽象符号完美融合，强调视觉象征意义的同时，突显文化的积淀性和历史的深厚感。设计师有意识地利用光、影、色构建空间的通透性，强调空间的装饰性和环境的隐喻性，打造出多元风貌并存的居住空间，传达人们对自由与个性的追求。

客厅：丝绒沙发与水晶吊灯的搭配将欧洲古典主义元素展示出来，而抽象艺术装饰画又将后现代主义的时尚文化理念彰显出来。这里有传统与现代的碰撞，更有自由与个性的全新风尚，经过设计师的孕育、融合、诠释和不断创新，创造出的居住环境经典而富有魅力。

餐厅：开放式的餐厅与客厅格调高度统一，欧式的浪

漫元素蕴藏于水晶灯、烛台和银质餐具之中，如此优雅的就餐环境非常适合一家人围坐在餐桌旁，共享天伦之乐。

书房： 以点、线、面来诠释后现代的抽象美，深沉的色调营造出安静的氛围，办公桌椅现代感十足，书架内隐藏的灯饰极具创意，不但补充了光源，更成为了一种装饰。

卧室： 卧室选用金褐色调作为主打色，体现成熟与稳重的气息，配以高贵与典雅的紫红色，增添了一种稳重的氛围。

The interior design of post-modern style can integrate many metaphorical visual symbols inside the works, creating some refreshing high taste experiences with everlasting charms. This project perfectly integrates luxurious elements of European aristocratic style and post-modern abstract symbols, emphasizing visual symbolic significance and highlighting cultural sedimentary deposits and historical profound feels. The designer consciously makes use of light, shadow and colors to construct the transparency of space, emphasizing the space decorative quality and environmental metaphorical nature. The designer creates a residential space combining multiple outlooks and displays people's pursuits of freedom and personalities.

Living Room: The combination of velvet sofa and crystal droplights presents the European classical elements, while the abstract artistic decorative paintings manifests the fashionable cultural concepts of post-modernism. There is the clash of traditions and modern elements, there is also the new fashion of freedom and personalities. Through fostering, integration, interpretation and constant innovation of designer, the residential environment acquires classical and constant charms.

Dining Hall: The open-style dining hall is highly integrated with the living room in tones and the European romantic elements are contained in the crystal lights, candlesticks and silver dining accessories. Such an elegant dining environment is quite fit for a family getting together around the dining table and enjoying the family happiness.

Study: The designer makes use spots, lines and surfaces to interpret the abstract beauty of post-modernism. Profound colors create some tranquil atmosphere and the office table and chairs are quite modern, the lighting accessories inside the book shelf is quite innovative, which not only supplement the light source, but also become a decoration.

Bedroom: The bedroom selects gold brown color as the tone color, representing mature and sedate atmosphere. The collocation of noble and elegant purple color strengthens the sedate atmosphere.

摩登样板间 III
新简欧

中洲中央公园 1-B02 实地样板间

Central Park 1-B02 Show Flat

设计公司：KSL 设计事务所
主设计师：林冠成
项目地点：广东深圳
项目面积：230 m²
主要材料：金影木、特殊玻璃、鹅毛金大理石、布板、皮革

Design Company: KSL DESIGN(HK) LTD.
Designer: Andy Lam
Project Location: Shenzhen of Guangdong Province
Project Area: 230 m²
Major Materials: Avodire, Special Glass, Marble, Fabric Swatch, Leather

雍容贵气的香槟金与浪漫尊贵的水晶紫相互组合，散发出欧式风格的典雅与华贵气息，简洁干练的现代设计语言简化了传统欧式的琐碎装饰，以明快而有气度的手法重新诠释了具有现代情调的空间。从精致的色彩搭配到点面相结合的空间构图，整个空间雅致，氛围柔美浪漫，使人体验贵族风范的同时，又能体验到闲适有度的生活情趣。

客厅：欧式风格的典雅气息蔓延到客厅的每个角落，灿烂的水晶灯和进口壁纸的色泽相互呼应，提升了空间的华贵感。欧式经典造型的丝绒面料沙发选用了紫罗兰色，与墙面的金棕色与咖色相互搭配，使色调极为舒展，空间氛围雍容有度、舒适自在。

书房：白色书架与黑色书桌形成对比，使书房的色调沉稳起来，可以带来宁静的感觉，从百叶窗射进来的温暖阳光让人心情大好。

玄关：入门玄关使人惊艳，椭圆形顶棚与纹饰地板首尾呼应，开门点题的欧式格调深入人心。

主卧：大量丝绒与绸缎面料的出现让卧室极为雍容，精致的花纹与细节装饰让人仿佛置身于古代宫廷。

Noble and aristocratic champagne gold is combined with romantic and noble crystal purple color, displaying some European style's elegant and magnificent atmosphere. The concise and brisk modern design language simplifies the trivial decorations of traditional European style, reinterpreting the space of modern charms with bright and grand approaches. From delicate color collocations to space composition combining spots and surfaces, the whole space is quite elegant, with soft and romantic atmosphere, which can make people experience the noble feel, while having the leisure life interests with proper extent.

Living Room: The elegant atmosphere of European style spreads to every corner of the living room, with brilliant crystal lights echoing the luster of imported wallpaper, uplifting the magnificent feel of the whole space. The velvet lining sofa of classical European format selects violet color, with gold brown color on the wall collocating

with coffee color, stretching the color tones to the extreme, while making the space atmosphere be appropriate, elegant and cozy.

Study: White bookshelf produces contrasts with the black table, making the color tone of study become sedate, bringing some tranquil sensations. And the warm sunshine getting inside through the window-blinds makes people feel great.

Hallway: The indoor hallway is quite amazing. The oval ceiling echoes the ornamentation board at the beginning and the end. The European tone at the hallway enjoys popular support.

Master Bedroom: The large amount of application of velvet and silk fabrics makes the bedroom appear magnificent. The delicate pattern and detail decorations make people seem to be inside an ancient palace.

中洲中央公园二期12-A01户型样板间
Central Park Phase 2, 12-A01 House Type Show Flat

设计公司：KSL设计事务所
主设计师：林冠成
项目地点：广东深圳
项目面积：130 m²
主要材料：水曲柳索漆、蓝金砂石、镜钢、皮板、木地板、壁纸板

Design Company: KSL DESIGN(HK) LTD.
Designer: Andy Lam
Project Location: Shenzhen of Guangdong Province
Project Area: 130 m²
Major Materials: Ash-Tree Lacquer, Blue and Gold Gravel, Mirror Steel, Leather Board, Wallpaper Board

如沐一曲悠扬的钢琴乐，温蔼的浪漫气息在这间新古典风格的样板间中缓缓弥漫开来，那是华美的水晶灯、精致的茶盏、清新的沙发烘托的美好氛围，更是清丽的湖蓝、雅致的赤紫、高贵的金色所营造的空间意蕴。本案从空间架构到软装配色，设计师都拿捏得十分精准，浪漫而不矫情，充满情调而不做作，设计师以细腻的笔触表达着一种美学品位与文化追求，更诠释出一种闲适恬淡的生活方式。

客厅： 白灰色打底，提升空间整体的清新感，配以浅棕墙面、金色沙发和纹饰地毯，使高贵的古典气息散发出来，点缀的赤紫色和湖蓝色则带来地中海的清新风范。配饰是古典的，设计手法却是现代的，简洁干练的设计让空间呈现自由、瑰丽的氛围。

餐厅：欧式古典餐桌椅的美感不言而喻，典雅之中掺入湖蓝色的清新感，配合精致的同色系餐具，雅致而不失活泼。

书房：白色的书房显得尤为宁静，墙面装饰的花纹使欧式古典风情散发出来，忙时可以处理公务，闲时可以阅读，一间书房就是一个世界。

卧室：顶棚装饰的圆形水晶灯衬托出华丽的氛围，墙壁的复古纹饰与整体格调保持一致，白色墙面与蓝色床品的搭配让卧室显得干净而舒适。

儿童房：粉红色是浪漫情调的最好代言，可爱的粉色卡通花纹布满整个墙面，丝绒材质的床头不仅柔软舒适，还提升了空间的华丽气氛。

All is like a melodious piano music. The romantic atmosphere is gently pervading inside the Neo-Classical style show flat. The grand crystal lights, delicate tea cup and fresh sofa set off some nice atmosphere. Those space connotations are also created by clear lake blue color, elegant purple color and noble gold color. From space construction to soft decoration color collocations, the designer masters quite accurately, being romantic but not hypocritical, full of charms. The designer makes use of refined brushworks to display some nice tastes and cultural pursuits, interpreting some leisurely and tranquil life style.

Living Room: The white gray color as foundation uplifts the fresh feel of the whole space, accompanied with light brown wall surface, gold sofa and ornaments carpet, bringing out the noble classical atmosphere. The decorated purple and lake blue colors create the fresh Mediterranean feel. The ornaments are classical, yet the design approach is quite modern. Concise and brisk design makes the whole space appear free and magnificent.

Dining Hall: The European classical dining table and chairs are quite aesthetic, with elegance incorporated with the fresh lake blue color, accompanied with same color dining tablewares, displaying some elegant yet dynamic atmosphere.

Study: The white study appears quite tranquil, and the pattern on the wall displays the European classical charms. You can deal with business here, as well as reading. A study can be a world.

Bedroom: The round crystal lights on the ceiling set off some magnificent atmosphere. The antique pattern of the wall is consistent in tone with the whole space. The combination of white wall and blue bedding accessories makes the bedroom appear tidy and comfortable.

Children's Room: Pink color is the best representation of romantic charms. Lovely pink cartoon pattern spreads on the whole wall. The velvet head of bed is not only soft and comfortable, but also uplifting the grand atmosphere of the whole space.

香港信和·中央广场示范单位1
Hong Kong Sino·Central Plaza Show Flat 1

摩登样板间 III
新简欧

设计公司：香港方黄建筑师事务所	Design Company: Hong Kong Fong Wong Architects & Associates
设 计 师：方峻	Designer: Noah Fong
项目地点：福建厦门	Project Location: Xiamen of Fujian Province
项目面积：449 m²	Project Area: 449 m²
主要材料：当地环保材料	Major Materials: Local Environmental Materials
摄 影 师：叶景旱	Photographer: Ye Jingxing
开发单位：香港信和地产	Project Developer: Hong Kong Sino Group

我们的设计灵感来自于瓦松灰。生于石质山坡和岩石上的瓦松,通常有一种很优雅的色泽,就是淡淡的灰白色,像是天鹅绒般的鼠灰或是冷冷的烟山色。因为这种颜色接近基本的协调色,不包含其他色彩,所以能很好地衬托出与之相搭配的其他颜色的明亮色泽。灰色且略带朦胧感的色调可以恰到好处地体现出一丝轻柔和神秘感。

灰色在一定程度上还代表了成熟及洒脱,也会给与人们舒适、平和的印象。

Our design inspirations come from herba orostachyos gray color. Herba orostachyos grows on stone hillside or rocks, usually having very elegant luster, i.e. light gray white color, of velvet gray color or cold mountain color. As that color is quite close to basic coordinating color, not including other colors, thus it can finely set off other colors' bright luster collocating with it. The gray color tone yet with some hazy feel can appropriately reflect some softness and mysterious feel.

The gray color can represent maturity and free spirit to some extent, while producing some cozy and peaceful impressions for people.

香港信和·中央广场示范单位2
Hong Kong Sino·Central Plaza Show Flat 2

设计公司：香港方黄建筑师事务所	Design Company: Hong Kong Fong Wong Architects & Associates
设 计 师：方峻	Designer: Noah Fong
项目地点：福建厦门	Project Location: Xiamen of Fujian Province
项目面积：449 m²	Project Area: 449 m²
主要材料：当地环保材料	Major Materials: Local Environmental Materials
摄 影 师：叶景星	Photographer: Ye Jingxing
开发单位：香港信和地产	Project Developer: Hong Kong Sino Group

我们的设计灵感来自于三色堇。

三色堇又名蝴蝶花,其色彩斑斓,在欧美地区颇受人们的喜爱。

些许纯白、几许微黄、几许紫色,三色堇的色调跳跃、生动:白色纯净清爽、黄色愉悦轻松、紫色神秘优雅。这几种色彩的搭配,相得益彰,很有层次感,并能营造愉快与华丽的氛围。

作为表现"思慕"情怀的花卉,三色堇充满生机,情趣盎然。

Our design inspiration comes from pansy.

Pansy is also called Butterfly Flower, with multicolors, very popular in Europe and America.

Some pure white, some light yellow and some purple colors. The color tone of pansy is dynamic and lively. White is pure and refreshing, yellow is pleasing and relaxing and purple is mysterious and elegant. The collocation of these colors brings out the best in each other, with great layer feel, while creating some pleasing and magnificent atmosphere.

As a flower representing sensations of "longings and admirations," pansy is full of vigor and interests.

MODERN SHOW FLAT III NEW JANE EUROPEAN

MODERN SHOW FLAT III NEW JANE EUROPEAN

上海徐汇园二号楼
Shanghai Xuhui Garden, Building No. 2

设计公司：玄武设计群	Design Company: Sherwood Design Group
公司网址：www.sherwood-inc.com	Company Website: www.sherwood-inc.com
设 计 师：黄书恒、欧阳毅	Designers: Huang Shuheng, Ouyang Yi
软装设计：玄武设计	Soft Decoration Design: Sherwood Design
项目地点：上海	Project Location: Shanghai
项目面积：172 m²	Project Area: 172 m²
主要材料：黑金峰、浅金峰、卡拉拉白、白水晶、黑白大理石马赛克、橡木染灰木地板、壁纸、黑檀木、雕刻板贴银箔、茶镜	Major Materials: Marble, Bianco Carrara Marble, Quartz, Black and White Marble Mosaic, Oak Gray Wood Floorboard, Wallpaper, Black Ebony, Engraving Plate of Silver Foil, Tawny Mirror
摄 影 师：王基守	Photographer: Wang Jishou

金银流线　隽书英式新风
古典英式风格，源自十八世纪的英国，设计师深刻领会到古典风格所展现的力与美，掌握对称与和谐的原则，以古典意境的门廊、繁复的线条天花、规律而对称的图腾，构筑稳重而大气的空间。

黄金比例　和谐深韵
设计者将家族的图腾，延展于主要厅区之间。玄关中，落地镜面延伸了视觉感，金色的几何纹理与端景相互映衬。客厅里，错落有致的矩形背墙设计，凸显素材的鲜明个性，创造虚实并置的观景层次，对照电视主墙的艳红色框景手法，以不同明度、纹理的石材层迭围塑，创造出饶有趣味的戏剧效果。

空间流转　视觉剪影
设计师保留英国绅士风的从容姿态，透过以人为本的思维探究餐厅情境的合理塑造。首先，藉由菱格形切割镜面形成视觉剪影，使人们成为空间的最佳主角，从天花板垂吊而下的璀璨水晶灯，正好与天花柔美的曲线谱出共鸣。

生活风景　俯拾即是
英国重视历史文化的传承，人们对于书房的重视程度不亚于开

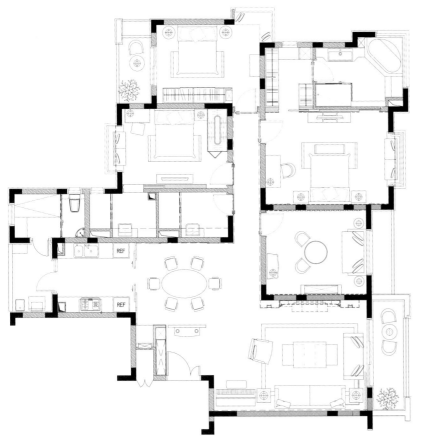

放厅区，布局通常以男主人为主。设计者依照生活习性一一归纳，理解书房是工作、品读之余，静心沉思的最佳场域。特别采用深沉色调铺陈周边墙面，以木作对应壁纸，一刚一柔的对比铺陈，增添了空间的宁静感。另外，尽可能简化机能摆设，将书柜融入墙面设计，随意摆放的书册与线条相映，不仅能节省空间，亦使画面呈现诗意氛围。

深具弹性的休憩氛围

秉承古典主义对称与和谐的原则，属于私人领域的卧房，以传递温暖为中心思想，细致窗纱对应纵横分明的图腾壁纸。同时，设计者精心铺陈冷暖色调，透过软装饰营造缤纷、欢乐的情调，使卧室空间在舒适氛围下，也体现出高度的艺术内涵。

Gold and silver filament lines describe the British new style.

Classical British style originates from England of 18th century. The designer profoundly finds out the strength and beauty presented by classical style, masters the principles of symmetry and harmony and constructs sedate and magnificent space with hallway of classical artistic conceptions, complicated lines ceiling and orderly and symmetrical totem.

Golden Proportion, Harmonious Rhymes

The designer extends the family totem inside the major

hall. For the hallway, the French mirror surface extends the visual feel, and the gold color geometric pattern and end scenes echo each other. For the living room, the well-proportioned rectangular wall design highlights the peculiar features of materials and creating visual layers of true and false lined together. Echoing the bright red framing approaches of the main TV wall, the scenes are created with stones of different brightness and patterns, producing dramatic effects of great interests.

Space on the Move, and Visual Silhouette

The designer retains the calm posture of British gentlemen style and explores the proper creation of dining hall environment from human-oriented thinking. First of all, the designer uses lattice format carving mirror surface to produce visual silhouette, maks people become the best protagonist of the space and the flamboyant crystal lights hanging on the ceiling produce resonance with the ceiling's soft curves.

Life sceneries are everywhere.

England focuses on inheritance of historical culture, and people pay as much attention to study as to open living room. The layout of study usually focuses on the host. The designer concludes the spaces according to life habits, and he understands that study is the best place for getting quiet down and meditation away from work and reading. The designer especially makes use of dark color tones to decorate the surrounding walls, with wood as the corresponding wallpaper. The powerful and soft contrasting layout adds to the space's tranquil feel. Other than that, the designer tries the best to simplify the mechanical setting, incorporating book cabinet into the wall design. The casually displayed books and lines bring out the best in each other, which not only saves spaces, but also displays some poetic atmosphere in the picture.

The flexible leisure atmosphere

The project inherits the principles of symmetry and harmony. The bedroom as private zone has conveying warmth as the main idea. The delicate window screening echoes the totem wallpaper of brisk horizontal and vertical pattern. At the same time, the designer carefully displays warm and cold color tone and represents great cultural connotations inside the bedroom's cozy atmosphere through soft decorations to create colorful and pleasing tones.

上海徐汇园三号楼
Shanghai Xuhui Garden, Building No. 3

设计公司：玄武设计
公司网址：www.sherwood-inc.com
设 计 师：黄书恒、欧阳毅
软装设计：玄武设计
项目面积：159 m²
主要材料：印度黑金、马赛克、壁纸、黑镜、木地板
摄 影 师：王基守

Design Company: Sherwood Design Group
Company Website: www.sherwood-inc.com
Designers: Huang Shuheng, Ouyang Yi
Soft Decoration Designer: Sherwood Design
Project Area: 159 m²
Major Materials: Indian Black Galaxy, Ariston Marble, Mosaic Tile, Wallpaper, Black Mirror, Wood Floorboard
Photographer: Wang Jishou

"真正的艺术品总是自然天成的，永远无法被我们所理解；只能观赏，只可感受；我们虽深受艺术影响，却浑然不知，更不消说要用言语来形容艺术的本质与贡献。"
——歌德

从心出发的设计

生活本质就是艺术，因为无从复制、无法重来。想要打造一个灵活的生活空间，设计必须由"心"出发，才能使空间尽善尽美。玄武设计师以满怀热情的设计态度，不断挑战设计的新高度，无所谓风格、不拘泥框架，于空间蓝图中展现独树一帜的创意。

光线视透 空间无限

L形的开放式公共区，透过大面窗景汲取自然光

线，呈现光影的动感，墙面横向的纹路与地毯的蜿蜒曲线，使空间鲜活起来。在极度扩张的巴洛克情境里，设计者运用经典的黑、白色，加上紫色与黄色的适度点缀，于流转的空间中塑造出一种静态的美感。

唯美的黄金分割律

本案的家具设计非常舒适，设计师藉由剪影的手法打造电视柜体，塑造出景中有景的剧院情节，横向延伸的线条依循文艺复兴的理性基础，衬托出空间的平衡感，客厅主墙两端大胆采用罗马柱设计，维持视觉稳定。

主厅区运用线帘、家具摆设作为隐性隔间，突显出场域的定位；餐厅具备不可思议的丰富性与自由度，天花的轮廓对应地面的繁复拼花、加上舒适的光源，使空间产生如舞台布景的华丽质感，体现丰富、自由的风格内涵。

"The authentic artistic objects are always naturally made and would never be understood by us. They can only be appreciated and sensed. Although we are deeply influenced by art, we do not perceive it, nor apply languages to describe the essence and sacrifice of art."

— Goethe

Design Arising from Heart

Life is art in essence, as it can not be reproduced nor restarted. If we want to create a dynamic living space, the design has to start from "heart", only in that way can the space be consummated. With enthusiastic design attitudes, Sherwood designers constantly challenge the new height in design, are not confined to styles nor frameworks, and they present some unique innovative ideas inside the space blue print.

Rays of Lights, Boundless Space

The L-shaped open public space draws natural light through the large window, presenting dynamic light and shadow. The horizontal lines on the wall and the zig-zag curves of the carpet make the space become alive. In the extremely expanded Baroque environment, the designer applies the classical black and white colors, dotted with proper purple and yellow colors, producing some quiet aesthetic feel inside the space on the move.

Aesthetic Golden Section

The furniture of this project is quite comfortable. The designer applies the approach of silhouette to produce the TV cabinet, building the theatre plots with views inside views. The horizontally extending lines follow the rational basis of

Renaissance, set off the balanced feel of the space and the two ends of dining hall's main wall applies the design of marble pillar to maintain visual stability.

The major hall area applies threads curtain and furniture as the inexplicit compartment, highlighting the orientation of regions. The dining hall attains inexplainable richness and freedom. The outline of ceiling echoes the complex parquet on the ground. Accompanied with comfortable light source, the space produces some magnificent feel like stage settings, displaying abundant and free style connotations.

摩登样板间 III
新简欧

未来之光样板间
Future Light, Show Flat

设计公司：玄武设计群	Design Company: Sherwood Design Group
公司网址：www.sherwood-inc.com	Company Website: www.sherwood-inc.com
设 计 师：黄书恒、欧阳毅、陈佳琪、蔡明宪	Designers: Huang Shuheng, Ouyang Yi, Chen Jiaqi, Cai Mingxian
软装设计：玄武设计	Soft Decoration Design: Sherwood Design
软装设计师：胡春惠、胡春梅	Soft Decoration Designers: Hu Chunhui, Hu Chunmei
项目地点：台湾新北	Project Location: Xinbei, Taiwan of China
项目面积：53 m²	Project Area: 53 m²
摄影师：王基守	Photographer: Wang Jishou
主要材料：金锋石、银狐石、蓝钻石、喷漆、雷射雕刻板、特殊壁纸	Major Materials: Jinfeng Stone, Silver Fox Stone, Blue Diamond Stone, Paint, Laser Engraving Plate, Special Wallpaper

桂冠礼赞 英伦风情

To see a world in a grain of sand
And a heaven in a wild flower,
Hold infinity in the palm of your hand
And eternity in an hour.

—— William Blake

从一粒细沙可以想到天地的宽广，从一朵野花可窥见世界的奥妙。只要用心观察体会，便会发现我们的双手掌握了无限可能，生活的每一瞬间都充满新奇与惊喜，只要掌握当下的每分每秒，创造出的价值便形同永恒。设计师汲取英国诗人布雷克动人诗篇的思想精髓，以英国维多利亚女皇时期繁复的艺术为基础，彻底颠覆现代风格所描绘的快餐文化似的空间感，酝酿出来自空间最深层的感动。

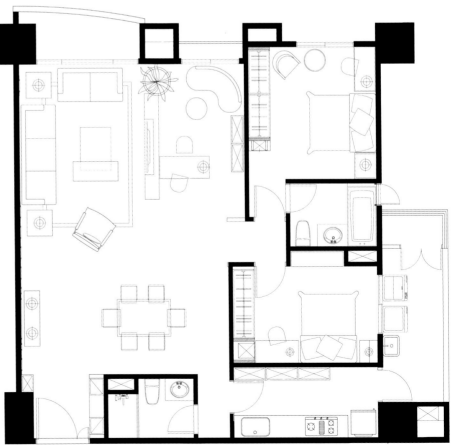

A Psalm of Laurel, British Charms
To see a world in a grain of sand
And a heaven in a wild flower,
Hold infinity in the palm of your hand
And eternity in an hour.
—— William Blake

From a grain of sand, we can think of the grandness of heaven and land and from a flower we can perceive the wonder of the world. If we observe sincerely, we can find that our two hands master limitless possibilities and every moment of life is full of wonders and surprises. If we master every minute at hand, we can create everlasting values. The designers draw the essence of English Poet Blake's touching poem and thoroughly overthrow the space feel described by modern style like fast-food culture, based on the heavy and complicated culture of English Victoria period, creating some affections from the very bottom of space.

MODERN SHOW FLAT III NEW JANE EUROPEAN

摩登样板间 III
新简欧

君悦黄金海岸样板示范单位
Junyue Grand Coast Show Flat

设计公司：台湾大易国际设计事业有限公司·邱春瑞设计师事务所
设 计 师：邱春瑞
项目地点：福建漳州
项目面积： 93 m²
主要材料：爵士白、黑白根、西班牙米黄、意大利木纹、银镜、木地板

Design Company: Taiwan Dayi International Design Industry Co. Ltd•Qiu Chunrui Interior Design
Designer: Qiu Chunrui
Project Location: Zhangzhou of Fujian Province
Project Area: 93 m²
Major Materials: Jazz White Marble, Black Marquina Marble, Spanish Beige Marble, Italian Wood Grain, Silver Mirror, Wood Floor

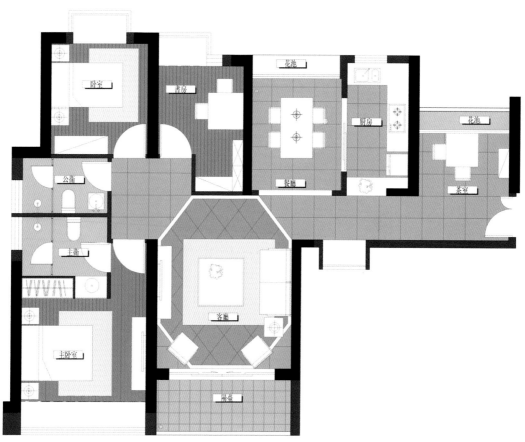

在这个空间中,设计师以柔和的奶白色为主基调、配合沉稳的咖色调的点缀,创造了一个雅致且有独特品位的生活场景。在简洁干净的界面下,不同质感和形态的家具及饰品既相互协调,又各自独立,碰撞出独特的时尚空间韵味。

在小配饰上,设计师也进行了精心的装点,精美的装饰画、雅致的餐具等,让人无时不在感叹生活的美好。在悠闲的午后,邀上三两好友享受着浓浓的咖啡,工作之余的劳累顿时一扫而光,生活的惬意就在于此。

For this space, the designer uses soft off-white color as the keynote and combines sedate coffee color tone to produce an elegant life scene of peculiar taste. Under the concise and clear interface, different texture and format furniture and ornaments are interconnected with each other, while being independent individually, producing peculiar fashionable space charms.

As for the little accessories, the designer makes some delicate ornaments. The exquisite decorative paintings, elegant tablewares and others make people be pleased with the beauty of life all the time. During the leisurely afternoon, you can invite several friends to enjoy the espresso. The exhaustion created by work would disappear all at once. And that is where the beauty of life lies.

MODERN SHOW FLAT III NEW JANE EUROPEAN 057

常州乐颐大厦某宅

One Residence of Changzhou Leyi Building

设计公司：上海廖易风建筑装饰工程有限公司
设计师：廖易风
项目地点：江苏常州
项目面积：240 m²

Design Company: Shanghai Liao Yifeng Architectural Decoration and Engineering Co., Ltd.
Designer: Liao Yifeng
Project Location: Changzhou of Jiangsu Province
Project Area: 240 m²

现在流行一种风格，叫"时尚新古典"。本案正是以这种风格为蓝本，将现在流行的欧式、现代风格相结合，提取各自的精华，满足现代人融品位与实用为一体的空间需求。

设计师采用现代的设计手法来营造古典的韵味，将欧式风情之家特有的典雅气质融入到空间中，从而迸发出一种独特的浪漫风情。层层递进的色彩与造型将空间延伸，让人可以感受到空间中蕴含的大家风范。整体色调淡雅、稳重且充满品位。

There is a style which is very popular nowadays which is called Fashion Neo-Classicism. This project has this style as the blueprint, combining popular European style and modern style, extracting the specific essence and meeting with modern people's space needs integrating tastes and practical needs.

The designer applies modern design approach to create classical charms, integrating the elegant temperament exclusive to European charms into the space, thus producing some peculiar romantic charms. The progressive colors and formats extend the space, making people feel the grand charms inside the space. The whole color tone is elegant, sedate and full of tastes.

摩登样板间 III
新简欧

长江路九号
Changjiang Road No. 9
No. 9

设计公司：东易日盛装饰集团
设 计 师：孟繁峰
项目面积：270 ㎡
主要材料：瓷砖、壁纸、实木多层地板、木纹玉

Design Company: Dong Yi Ri Sheng Home Decoration Group Co., Ltd.
Designer: Meng Fangeng
Project Area: 270 m²
Major Materials: Ceramic Tile, Wallpaper, Solid Wood Multilayered Floor, Wood Grain Jade

本案业主比较低调，温文尔雅，设计师觉得家的气质就应该代表主人的气质，想用优雅的设计语言来诠释这个作品。

户型弊端剖析

本案为270㎡的位于市中心区的高档公寓，这个户型在该小区仅有4套，也是最大的4套，实质上是160㎡公寓的复加，因为2梯一户的设计和一楼中奢华的会客大厅使得空间得房率极低。客厅挑空区域还需要选辟出通往二楼的通道，因此空间的"开放"成为关键。

方案调整意见

风格定位：根据居室的实际情况及业主的喜好，设计师建议将本案设计成带有古典意味却不繁冗厚重的新装饰主义风格，这种风格古典且不失现代感，明快优雅。

结构调整: 楼梯位置的选择对于别墅和复式户型特别重要，本案中，直跑小拐弯的楼梯设计使得客厅空间得到了最大化的利用，同时使得楼上的起居室更加开阔。客户有大量的玉器与翡翠收藏，希望在空间中得到展示，因此，每个功能区的墙面都采用了通透的处理方式，既让空间隔而未绝，也使收藏得到了展示。狭小的厨房不能让使用功能得到充分的发挥，因此将西厨进行了拓展，承担了厨房部分的收纳和操作功能。一层主要承载会客与客住的问题，二楼基本成为主人自己的独立空间，同时，设置了一个儿童空间，以便将来的需要。

关于色彩层次

设计师选用了帕拉迪奥珍珠白木门，希望用浅色勾勒出整个空间的层次，帕拉迪奥木门线条优雅，奠定了所配套的石膏线、踢脚线等装饰线条的基调。带珠光的浅香槟色、淡紫色、银灰色、卡布奇诺、淡绿色诠释了奢华、内敛、优雅的室内环境，也满足了每一个空间所有者的需求。通透的水晶灯使得空间更加透彻纯净，卡其色、珠光白的家具在这些色彩的映衬下和谐而优雅，咖啡色和淡米色微晶石的搭配让空间轮廓清晰，通透性再次得以提升，高光的烤漆面和奢华的蟒皮纹理与墙面、地面的材料搭配协调，丝质同色绣花窗帘继续将这一轻盈、纯净、优雅的主题延续。

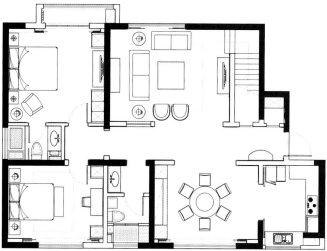
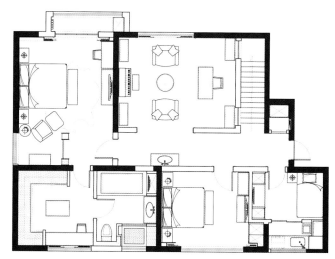

The property owner of this project is quite low-profile, gentle and cultivated. The designer believes that a home shall represent the temperament of the property owner and he wants to interpret this work with elegant design language.

Analysis on the Drawbacks of House Type

This is a high-end apartment of 270 m² located in the urban center. There are only 4 suits of this kind of house type in this residence and this project is the biggest one, as being in fact the duplex of 160 m² apartment. Due to the design of two staircases for one apartment and the luxury reception hall on the first floor, the usable area is quite small. There needs to be a corridor leading to the second floor for the living room's high space, thus it is the key point to "open" the space.

Project Adjustment Opinions

Style Orientation: According to the actual conditions of the residence and the host's likes, the designer suggests to design this project

with New Decoration Style with classical feel but not profound nor burdensome. This style is classical but with modern feel, displaying brisk and elegant atmosphere.

Structure Adjustment: The selection towards the location of staircase is quite important towards the house type of villa and duplex building. For this project, the staircase design of straight line and little bend maximize the application of the living room space, while making the living room upstairs more expansive. The property owner has a large collection of jade and emerald, which he hopes to present inside the space. Thus, the wall of each functional area applies transparent treatment, which separates the space yet with some connections, which can present the collections. The narrow kitchen can not make full use of the use functions, thus the western style kitchen is expanded, which assumes part of the kitchen's storage and operational functions. The first floor mainly undertakes the functions of receiving guests and guest accommodation. The second floor basically becomes the property owner's exclusive

independent space, together with children's space for future needs.

On Color Gradation

The designer selects Palladio pearl white timber doors to delineate the layers of the whole space with light color. This kind of doors have elegant lines, which establish the tone of decorative lines such as plaster lining, skirting lining, etc. The pearl luster light champagne color, light purple color, silver gray, cappuccino color and light green interpret the luxurious, restrained and elegant interior environment, while meeting with the requirements of each space user. The transparent crystal lights make the space appear pure and transparent, the furniture of khaki and pearl white color appear quite harmonious and elegant set off by those colors. The collocation of coffee color and light beige microlite allows the space to have clear outline, thus the transparency is uplifted further. The stoving varnish surface and the luxurious python skin pattern is in harmony with the materials of wall and ground. The same color silk embroider curtain further continues this lithe, pure and elegant theme.

黑白森林
Black and White Forest

设计公司：易百装饰（新加坡）国际有限公司
设 计 师：冯易近
项目面积：155 m²
主要材料：微晶石、饰面板、软包、壁纸等

Design Company: EBEY Decoration (Singapore) International Co., Ltd.
Designer: Feng Yijin
Project Area: 155 m²
Major Materials: Microlite, Veneer, Soft Roll, Wallpaper

业主背景：业主非常注重设计品质，对设计有着较高的要求。工作性质要求他要长期到外地出差，是一位事业有成的成功企业家，喜欢黑白色调。

设计思想：以黑白为主的设计主题成为本案的中心，这种色调代表了业主稳重、坚毅的性格特征，也体现出空间的经典与永恒感。黑色从面积最大的电视墙开始，一直延续到书房，再以白色的屏风和主体沙发来平衡黑色调的沉重感。

原书房的位置是卫生间，为将公共空间的面积扩大，将它改为客厅的共享空间，使空间规划更为合理。

地面采用白色砖进行大面积的铺设,再以不同宽度的黑色进行缓冲,不同比例的黑白铺设使得氛围非常融洽。

在空间中你会发现,纯色的白与彻底的黑相互映衬在一起,两种极端的色彩搭配在一起,使居室反而产生了一种独特的魅力。

Property Owner's Background

The property owner quite stresses on design quality and has comparatively high requirements for design. As requested by work nature, he needs to go on business trip for a long time. As a successful entrepreneur, he likes black and white color tone.

Design Concept: The design theme focusing on black and white becomes

the core of this project, which represents the property owner's prudent, firm and persistent characteristics, while displaying the space's classical and everlasting feel. The black color starts from the TV background wall with the biggest area, continues all the way to the study, and balances the heaviness of black color tone with white screen and major sofa.

The area of the original study is a washroom. And it was changed into the shared space of the living room in order to expand the area of the aforementioned shared space, making the space planning more proper.

The ground applies large area pavement with white tile, applying black color of different breadth for buffering. Black and white pavements of different proportions create harmonious atmosphere.

Inside the space, you may find that the pure white and thorough black color set off each other. The combination of these two extreme colors make the residence display some peculiar charms.

摩登样板间 III
新简欧

苏宁睿城某住宅
One Residence of Suning Ruicheng

设计公司：大品装饰 | Dolong 设计
施工单位：大品专业施工
项目地点：江苏南京
项目面积：140 m²
主要材料：大理石砖、皮革硬包、不锈钢线条、雕花灰镜、软木地板等

Design Company: Dapin Decoration | Dolong Design
Construction Company: Dapin Professional Construction
Project Location: Nanjing of Jiangsu Province
Project Area: 140 m²
Major Materials: Marble Bricks, Leather Hard Roll, Stainless Steel Lining, Carving Gray Mirror, Cork Floor

本案坐落于南京河西苏宁睿城小区，业主对完美生活理念有着独特的追求。

设计师用后现代的新装饰主义手法，运用简约的线条，比例完美的转折与分割，使室内空间处处充满个性的符号。客厅与餐厅呈开敞式，相互独立又相互统一，灰镜与亚光的硬包穿插在墙面上，相同的地面材质使空间散发出尊贵、大气的格调。后期搭配质感纯粹的家具，造型感强烈，散发出奢华欧式的高贵与典雅气息。

主卧室及卫生间则针对使用者的要求及喜好分别赋予不同的功能性，却依然能够取得与整体协调一致的良好效果。

This project is located inside Nanjing Hexi District's Suning Ruicheng Residential Community. The property owner has peculiar pursuits for the concept of perfect life.

The designer applies post-modern New Decoration approaches, concise lines, turning and segmentation of perfect proportion make the interior space be filled with unique symbols. The living room and dining hall are open style, mutually independent while being connected with each other. Gray mirror and hard rolls of dark color are interlaced on the wall. The same wall materials make the space send out some aristocratic and magnificent tones. For the latter stage, there are furniture of pure texture, with profound format feel, which sends out some noble and elegant atmosphere of luxury European style.

The master bedroom and washing room are entrusted with different functional features according to requirements and likes of users, but with fine effects consistent with the whole style.

中央原著某住宅
One Residence of Central Villa

设计公司：深圳市伊派室内设计有限公司
项目地点：广东深圳
项目面积：150 m²
主要材料：镜面、布艺、石材等

Design Company: Shenzhen Yipai Decoration Co., Ltd.
Project Location: Shenzhen of Guangdong Province
Project Area: 150 m²
Major Materials: Mirror, cloth, stone etc.

在本案中，设计师将现代与新古典的元素相结合，为业主营造一个高雅的生活环境，使他们体验到一种高品质的生活方式，表现出业主不俗的品位。设计师将欧式风格的经典元素用简化的新古典主义语言来表达高贵与典雅的风格，彰显贵族气质。

在色调上以沉稳的咖色为主，大面积柔和的淡黄色为背景，局部点缀白色、黑色，为生活空间注入一种怀旧的气息。简易的家具与色彩明快的布艺饰品，为空间增添了温馨的气息，彰显出空间内敛的格调，雅致而生动的摆设表现了主人别具慧眼的艺术修养。

For this project, the designer combines modern and classical elements, creating an elegant living environment for the property owners, and making them experience the high quality life style, displaying the property owner's uncommon taste. The designer uses classical elements of European style and simplified Neo-Classical language to display noble and elegant style, manifesting aristocratic temperament.

The color tone focuses on sedate brown color, with large area soft faint yellow as the background and black and white ornaments in some parts, instilling some nostalgic atmosphere inside the living space. The simple furniture and fabrics ornaments of bright colors add some warm atmosphere for the space, displaying the restrained tone of the space. The elegant and dynamic decorations represent the peculiar artistic attainments of the property owner.

摩登样板间 III 新简欧

世贸中心邬宅
World Trade Center Wu's Mansion

设计公司：宁波江北 UI 室内设计有限公司
设 计 师：陈显贵
项目地点：浙江宁波
项目面积：160 m²
摄 影 师：刘鹰

Design Company: Ningbo Jiangbei UI Interior Design Co., Ltd.
Designer: Chen Xiangui
Project Location: Ningbo of Zhejiang Province
Project Area: 160 m²
Photographer: Liu Ying

本案业主是一位成熟的服装设计师。在美国纽约生活的经历，使她有着独特的审美观念，喜爱游走在东方文化和西方文化的碰撞中，东方文化的素雅韵味与西方的时尚风采相互碰撞，使她呈现出浪漫且具有内涵的独特气质。

本案的设计依据女主人的性格及喜好量身定做。玄关处的中式手绘端景柜与欧式定制屏风完美结合，展现出主人对中西文化融合的喜爱，更值得一提的是客餐厅区域的门套和玄关区墙面大胆的撞色，再经过后期软装饰的搭配，就显得更加时尚、靓丽。客厅背景墙上的定制仿古镜、花鸟写意手绘壁纸、沉稳内敛的皮质沙发，使得空间中弥漫着成熟稳重而又温馨的格调。

外奔波多年的心，重又归于平静。儿童房活泼靓丽的竖条纹壁纸，定制的各式抱枕，体现了儿童活泼好动的性格特点，也体现出浓浓的亲情。

The property owner is a mature costume designer. The experiences of living in US New York allow her to have peculiar aesthetic viewpoints, and she enjoys moving among the clashes of oriental and western culture. The clashes between the elegant charms of oriental culture and the western fashion demeanors make her display romantic and peculiar temperament with profound connotations.

The design of this project is custom-made based on the hostess' characteristics and likes. The Chinese hand-painting end-view cabinet at the hallway is integrated perfectly with the European custom-made screen, demonstrating the owner's likes for integration of eastern and western culture. What's more, the bold contrasting colors between guest dining room's doorframe and the wall of hallway appear more fashionable and gorgeous, through accommodations of latter-stage soft decorations. The custom-made archaized mirror on the background wall of living room, wallpaper of flowers and birds painting in traditional Chinese style, the sedate leather sofa and others

make the space full of mature, dignified and warm styles. The sedate warm gray tone of the master's bedroom produces a warm and cozy leisure environment, soothing the heart for too long busy rushing about in the outside world, now

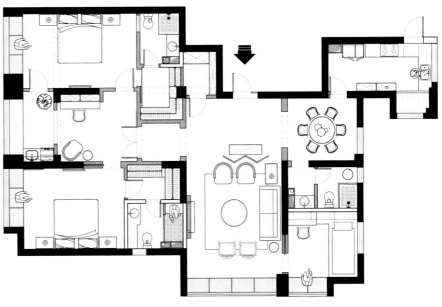

everything is back to tranquility. The dynamic vertical stripe wallpaper inside the children's room and the custom-made cushions of various style represent the vigorous characteristics of children, while with intensive family affections.

锦华园山景水岸样板间
Jinhua Garden Mountain View, Show Flat

设计公司：大墅尚品·由伟壮设计
设 计 师：李跃、黄杰
软装设计：翁布里亚专业软装机构
施工单位：大墅施工
项目地点：江苏南通
项目面积：135 m²
主要材料：仿大理石砖、橡木地板、壁纸、护墙板等

Design Company: Dashu Shangpin•Zhuang Design
Designers: Li Yue, Huang Jie
Soft Decoration Designer: Umbria Professional Soft Decoration Institution
Construction Company: Dashu Construction
Project Location: Nantong of Jiangsu Province
Project Area: 135 m²
Major Materials: Simulated Marble Bricks, Oak Floor, Wallpaper, Wainscot Board

MODERN SHOW FLAT III NEW JANE EUROPEAN

本案以欧式风格为主题，在设计上追求空间变化的连续性和形体变化的层次感，整体上简约的线条和几何图案与局部的相对繁琐形成鲜明对比，带给我们不一样的视觉及触觉享受。

色彩上以经典的米黄色与白色为主，再加上银镜、雕花镜、软包、壁纸等材料的贯穿融合，既突显空间的凹凸感，又带有欧式的优雅高贵感，符合中国人的审美价值。

布艺的沙发有丝绒般的质感及流畅的木质曲线，将传统欧式家居的奢华与现代家居的实用性完美地结合。还有精美的油画，制作精良的雕塑工艺品，都是打造欧式风格不可缺少的元素。

This project has European style as the theme and aspires for continuation of space changes and layers in format variations in design. The whole concise lines and geometric graphics produce sharp contrast with the relative complicated parts, creating some different visual and touch enjoyments for people.

The color system focuses on classical beige and white colors, together with the integrations of materials such as silver mirror, carving mirror, soft roll, wallpaper, etc., which not only highlight the concave and convex feel of the space, but also possess some European elegant noble feel, in accordance with Chinese people's aesthetic attitude.

The cloth sofa has velvet texture and smooth wood curves, perfectly combining the luxury of European furnishing and practical nature of modern furnishing. Exquisite oil paintings and excellent sculpture artworks are also indispensable elements in creating European style.

摩登样板间 III
新简欧

大奢至尚
Grand Luxury

设计公司：大品装饰 | Dolong 设计
施工单位：大品专业施工
项目地点：江苏南京
项目面积：195 m²
主要材料：大理石、皮革硬包、黑钛不锈钢线条、玉石砖等
摄　　影：金啸文空间摄影

本案位于南京保利香槟小区，经过设计师的精心布置及对空间和格局的独到把握，整个空间显得温馨而浪漫。黑与白是这里的主基调，很少有设计师会把这样的颜色用在新古典的设计中，在本案中，设计师经过与业主的沟通，将这样的想法赋诸于空间，才成就了空间的独一无二。

Design Company: Dapin Decoration | Dolong Design
Construction Company: Dapin Professional Construction
Project Location: Nanjing of Jiangsu Province
Project Area: 195 m²
Major Materials: Marble, Leather Hard Roll, Black Titanium Stainless Steel Lining, Jade Brick
Photographer: Jin Xiaowen Space Photography

客、餐厅运用简练的天花线条和大理石，再选用香槟色的硬包材质、高反光度的黑钛不锈钢、色彩凝重的家具来强化作品的特色。银色与香槟色的运用较为大胆，从灯具到壁画框边、家具镶饰，似流水般的银色给人以气质高雅的感觉。银色和香槟色成为黑、白两色的过渡，柔化了黑与白的硬朗。

This project is located inside Nanjing Poly Champagne Residence. Through the designer's careful setting and peculiar mastering towards space and layout, the whole space appears warm and romantic. Black and white are the keynotes here and rarely are there designers who would apply such colors in Neo-Classical design. For this project, through negotiations with property owner, the designer applies such ideas to the space and achieves unique quality of the space.

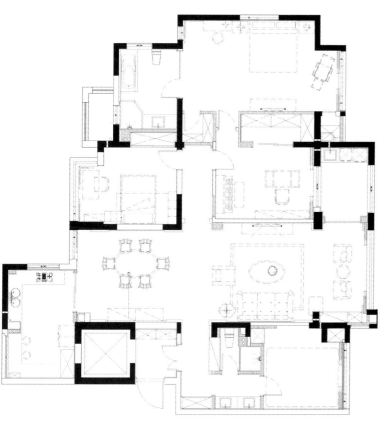

The living room and dining hall apply concise ceiling lines and marble, together with champagne color hard roll materials, black titanium stainless steel of high glossiness and furniture with dark colors to strengthen the features of the space. The designer is quite audacious in the application of silver color and champagne color, from lighting accessories, to mural framing and furniture lining, which flowing-water like silver color leaves people with elegant sensations. The silver and champagne colors become the traditional colors between black and white, softening the hardness of these colors.

MODERN SHOW FLAT III NEW JANE EUROPEAN 109

百悦天鹅湖样板间
Swan Lake Property, Show Flat

设计公司：成都多维设计师事务所
设 计 师：张晓莹
项目地点：四川成都
项目面积：110 m²
主要材料：白色大理石、米黄云石、米黄洞石、云石马赛克、马赛克、灰镜、木地板等

Design Company: Chengdu DODOV Design Studio
Designer: Zhang Xiaoying
Project Location: Chengdu of Sichuan Province
Project Area: 110 m²
Major Materials: White Marble, Beige Marble, Beige Travertine, Marble Mosaic, Mosaic Tile, Gray Mirror, Wood Floorboard

本案的欧式新古典风格与传统的古典风格相比，不再呈现无比厚重、庞大的气质，而是向轻盈、小巧发展，充满时尚感。本案采用各种颜色的大理石搭配马赛克、玻璃等，充满小资情调，高贵而不浮华。

欧式新古典的家具一反纯粹古典主义的厚重、沉闷，体量不再一味求大求重，造型也不再繁复；材质不再以木质、皮质为主，而是多种材料并行，营造出一个轻灵、透亮的空间。

总的来说，新古典风格属于一种大众风格，无论是偏古典还是偏现代，都可以达到简约精致的效果，既能显得时尚，也能做到温馨。

Compared with the traditional classical style, the European Neo-Classical style of this project does not present the much too profound or grand temperament, orienting towards light, graceful, small and exquisite style, full of fashion feel. This project applies the combination of marble of various colors and mosaic tile, glass, etc., full of petty bourgeoisie appeals, being noble but not flamboyant.

Different from pure classicism of heavy and dull outlook, the furniture of European Neo-Classical furniture does not seek for heavy mass, nor complicated formats. The materials would no longer focus on wood or leather, but with multiple materials, producing a lively and bright space.

In general, Neo-Classical style is a public style. Be it classical or modern in style, they can always attain the concise and delicate effects, appearing fashionable and warm at the same time.

摩登样板间 III
新简欧

常州莱蒙城别墅
Changzhou Laimeng Town Villa

设计公司：上海廖易风建筑装饰工程有限公司
设 计 师：廖易风、马利华、肖飞生
项目地点：江苏常州
项目面积：400 m²
主要材料：米黄大理石、软包墙、进口壁纸、波斯块毯、进口方块地板、白色硝基漆等

Design Company: Shanghai Liao Yifeng Architectural Decoration and Engineering Co., Ltd.
Designers: Liao Yifeng, Ma Lihua, Xiao Feisheng
Project Location: Changzhou of Jiangsu Province
Project Area: 400 m²
Major Materials: Beige Marble, Soft Roll Wall, Imported Wallpaper, Persia Block Blanket, Imported Square Floor, White Nitro-lacquer

在本案中，设计师力图表现出优雅与简洁相结合的精致感，以求在繁忙嘈杂的都市，营造一个可以完全放松、回归自然的生活场所。

本案是一套高贵典雅的新古典风格的设计作品，立体感强的大理石墙面装饰、丰富的颜色、精美的家具、多彩的壁纸、古朴的壁炉、木结构的楼梯、丰富的饰品等都是新古典风格的完美演绎。设计上线条流畅、尺度亲切适宜，呈现出经典、朴素、简约的生活之美。它所带来的质朴、舒缓的家庭氛围，满足了不同年龄层的需求，灰色调的装点，使一切都变得那么和谐而融洽。

For this project, the designer tries the best to display the delicate feel of the combination of elegance and conciseness, thus creating a living sphere where one can get fully relaxed and return to nature in this busy and noisy city.

This noble and elegant design work has Neo-Classical style. The marble wall decorations of intensive three-dimensional feel, rich colors, delicate furniture, colorful wallpaper, primitive furnace, abundant ornaments and wooden staircase are all perfect interpretations of Neo-Classical style. The design has smooth lines and cordial and intimate scale, presenting the classical, primitive and concise life beauty. The primitive and relieving home atmosphere that the design creates meet with the requirements of different age groups. The ornaments of gray color tone make everything appear quite harmonious.

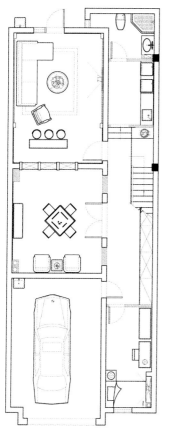
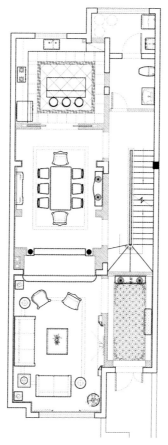
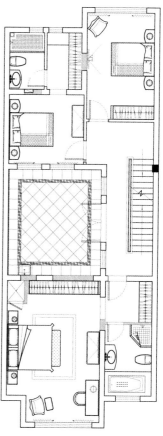

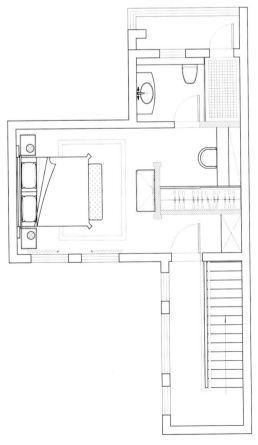

MODERN SHOW FLAT III NEW JANE EUROPEAN

摩登样板间 III
新简欧

城市时代 T1 样板间
City Time, T1 Show Flat

设计公司：深圳市帝凯室内设计有限公司
设 计 师：徐树仁
项目地点：广东惠州
项目面积：134 m²
主要材料：雅士白大理石、黑白根大理石、车边银镜、黑镜钢、珠光白漆饰面、钻石绒硬包、艺术壁纸

Design Company: Shenzhen Dikai Interior Design Co., Ltd.
Designer: Xu Shuren
Project Location: Huizhou of Guangdong Province
Project Area: 134 m²
Major Materials: Jazz White Marble, Black Marquina Marble, Silver Mirror, Black Mirror Steel, White Paint Veneer, Diamond Velvet Hard Roll, Artistic Wallpaper

本案设计概念是对于具有高贵气质与摄人风采的女性形象的想象，风格奢华而不奢靡，贵气而不张扬，简化的古典线条，带着一种悠闲舒适感。空间富有的钻石绒硬包及镜面对空间的延伸，让空间的质感细腻地呈现出了别样的奢华度。大面积的珠光白漆饰面增加了空间的温婉之气，极富心思的家具配饰，隐约的显露了她的内在美。设计师想让人感觉样板空间娴静舒适，让空间中高贵优雅、倾城倾国之气弥散开来。

The design concept of this project is oriented for female image of noble temperament and breathtaking glamours, with luxurious but not extravagant style, aristocratic but not showy and the simplified classical lines carry some leisure and cozy feel. The abundant diamond velvet hard rolls inside the space and the extension of mirror surface to the space display the texture of space with some peculiar luxurious quality. The large area of white paint veneer adds to the gentle feel of the space and the ingeniously selected furniture ornaments subtly display her inner beauty. The designer wants to make the show flat space appear tranquil and cozy, noble and elegant, with exceedingly beautiful appearance.

简·绎
Conciseness·Presentation

设 计 师：黄耀国
项目面积：135 m²
摄 影 师：施凯

Designer: Huang Yaoguo
Project Area: 135 m²
Photographer: Shi Kai

本案摒弃传统欧式风格繁琐的奢华造型，设计师通过对各种简洁的装饰元素的应用，延续了时尚的人文生活态度。

本案业主是一位从事服装生意的时尚女性，个性随和，乐于享受生活。考虑到业主的实际需要，主卧设计了独立的更衣间，扩大了主卧空间的同时，也增加了采光效果。

在材质运用上，设计师大量使用天然爵士白大理石，使整个空间灵动起来。白色木门、白色餐桌等白色元素的大范围使用，为整个空间奠定了简洁明亮的基调，给人一种犹如置身云端的视觉感受。另外，咖色的硬包造型，经典的大马士革图案的马赛克，在简洁的基调上延续出了几分稳重与端庄。

MODERN SHOW FLAT III NEW JANE EUROPEAN

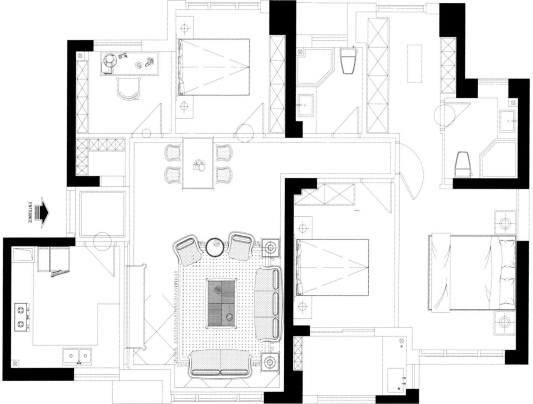

This project abandons the tedious luxury format in traditional European style, through the application of various concise decorative elements, continuing the fashionable human living attitudes.

The property owner of this project is a stylish woman engaged in costume business, with easy-going personalities and taking delight in enjoying life. Considering the actual needs of property owner, the master bedroom has an independent dressing room, while expanding the space of master bedroom, adding to lighting effects.

As for the application of materials, the designer makes a lot use of natural Jazz White Marble, making the whole space become dynamic. The extensive application of white elements such as white door and white dining table sets a concise and bright tone for the whole space, making people feel like being above the clouds. Other than that, the coffee color hard roll format and the classical mosaic tiles of Damascus graphics continue the sedate and solemn tone based on the concise tones.

桂丹颐景园高层样板间
Guidan Yijingyuan High-rise Show Flat

设计公司：SDD 上达国际
　　　　　深圳市上达建筑装饰设计有限公司
设 计 师：刘海涛
项目地点：广东佛山
项目面积：140 m²
主要材料：帝皇金大理石、浅啡网大理石、米黄石、金蜘蛛大理石、
　　　　　仿古砖

Design Company: SDD Design International
Shenzhen SDD Architectural Decoration and Design Co., Ltd.
Designer: Liu Haitao
Project Location: Foshan of Guangdong Province
Project Area: 140 m²
Major Materials: Emperor Gold Marble, Light Coffee Color Marble, Beige Stone, Spider Golden Marble, Archaized Tile

"时光流转,岁月鎏金",只有带着岁月流淌的痕迹,才能激起内心深处对"家"这个概念的深切体会。

本案在空间格局的安排上,注重实用性。140 ㎡的空间设计,既要满足三代同堂的需求,又要不显局促,是考验设计师空间把握能力的关键。所以,设计师在电视背景墙周围应用银镜来装饰,餐厅区背景墙也采用不规则的银镜拼接,就是为了更好地从视觉上使空间延伸、放大。

本案以白色为主色调,运用香槟色花纹壁纸、浅咖和深咖硬包来烘托室内的奢华效果。天花墙面线条的处理,简单明快,既不缺乏欧式的华贵与典雅,又符合现代人的审美情趣。

"Time and tide wait for no one." Only the traits with the flowing time can arouse from the bottom of people's heart deep understandings towards the concept of "home".

For this project, the space layout focuses on practicality. The space design of 140 m² should not only meet with the requirements of three generations under one roof, but also not appear much too crowded. That is the key in testing the designer's space mastering capabilities. Thus, the

MODERN SHOW FLAT III NEW JANE EUROPEAN

designer decorated the surrounding area of TV background wall with silver mirror, while the background wall of dining hall is also connected with irregular silver mirror, for better visually extending the space.

This project has white as the tone color and applies champagne floral pattern wallpaper and light coffee, dark coffee hard rolls to set off the interior luxurious effects. The treatment towards the ceiling wall lines is simple and brisk, being not only with European aristocracy and elegance, but also in accordance with modern people's aesthetic interests.

宏侨凯旋名门一期15#样板间

Hongqiao Triumphal Arch, Phase 1, 15# Show Flat

设计公司：福建国广一叶建筑装饰设计工程有限公司	Design Company: Fujian Guoguangyiye Architectural Decoration and Design Engineering Co., Ltd.
设计师：庄锦星	Designer: Zhuang Jinxing
设计审定：叶斌	Project Examiner: Ye Bin
项目地点：湖南株洲	Project Location: Zhuzhou of Hunan Province
项目面积：80 m²	Project Area: 80 m²
主要材料：大理石、钨钢镜面不锈钢、墙纸、镜面、软包等	Major Materials: Marble, Tungsten Steel Mirror Surface Stainless Steel, Wallpaper, Mirror Surface, Soft Roll
摄影师：施凯	Photographer: Shi Kai

简约欧式风格沿袭古典欧式风格的主元素，融入了现代的生活理念。有的不只是豪华大气，更多的是惬意与浪漫。通过完美的典线，精益求精的细节处理，带给家人不尽的舒适感。古典欧式风格线条复杂、色彩低沉，而简欧风格则在古典欧式风格的基础上，以简约的线条代替复杂的花纹，采用更为明快清新的颜色，既保留了古典欧式的典雅与豪华，又与现代生活所追求的悠闲与舒适相适应。另外，欧式风格最适用于大面积的房子，若空间太小，不但无法展现其风格气势。

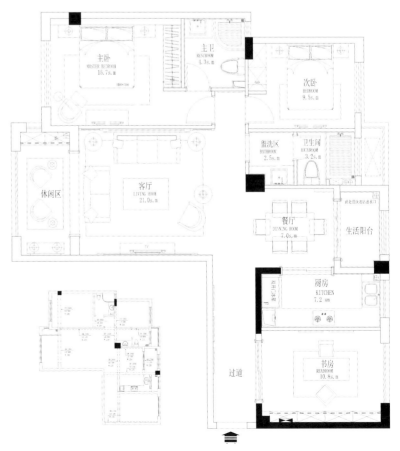

MODERN SHOW FLAT III NEW JANE EUROPEAN

Jane European style inherits the major elements of classical European style, while incorporating the modern life concepts. What it owns is not only luxury and magnificence, but also leisure and romance. The perfect curves and exquisite details treatment create limitless comfort feel for the family members. Classical European style has complex lines and dark colors. Based on classical European style, Jane European style replaces complicated patterns with concise lines, applying more bright and fresh colors, while maintaining the elegance and luxury of classical European style, being in accordance with the leisure and comfort that modern life pursues. Other than that, European style is quite fit for houses with large space. Small space would be hard to display its style and momentum.

欧式风情遇上现代简约

When European Charms Meet with Modern Conciseness

设计公司：深圳壹叁壹叁装饰有限公司
设 计 师：温旭武
项目地点：广东惠州
项目面积：126 m²

Design Company: Shenzhen 1313 Decoration Co., Ltd.
Designer: Wen Xuwu
Project Location: Huizhou of Guangdong Province
Project Area: 126 m²

本案通过一系列极富欧洲风情的装饰元素，营造出一种典雅、自然的生活气质，闲适、浪漫的空间情调是本案的主题。

设计师运用欧洲风格的软装配饰，比如家具、印花地毯、薄纱窗帘等，营造出一种古朴的欧式风情，与现代简约的空间形态完美配合，体现了现代生活所需要的简约和实用，又兼具传统欧式风格，富有朝气、韵味十足。

在整体设计上，本案空间布局开阔而大气，通透而时尚感十足。色彩搭配合理，给人以明朗舒缓的心理感受。不得不说，这是一个充满情调的个性化十足的空间。

Through a series of decorative elements full of European charms, this project produces some elegant and natural life temperament and the cozy and romantic space charms are the theme of this project.

The designer applies soft decorations of European style, such as furniture, printed carpet and chiffon curtain to create some primitive European charms, which are in perfect combination with modern concise space format. That not only represents the conciseness and practicality that modern life requires, but also possesses European style, full of vitality and artistic conceptions.

For the whole design, the space layout of this project is broad and expansive, being transparent and very fashionable. The proper color collocations leave people with bright and soothing psychological impressions. We have to say that this is a peculiar space full of emotional appeals.

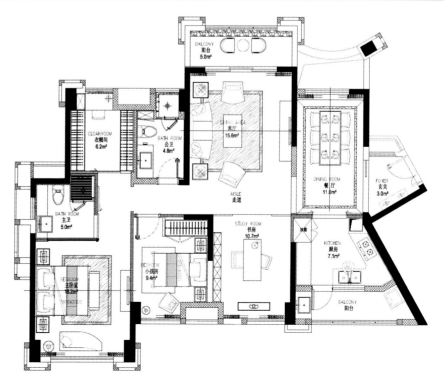

MODERN SHOW FLAT III NEW JANE EUROPEAN

恒信·中央公园9C 户型样板间
Hengxin·Central Park 9C House Type, Show Flat

设计公司：深圳市盘石室内设计有限公司 吴文粒设计师事务所	Design Company: Shenzhen Huge Rock Interior Design Co., Ltd./ Wu Wenli Design Firm
设 计 师：吴文粒、陆伟英	Designers: Wu Wenli, Lu Weiying
项目地点：湖北宜昌	Project Location: Yichang of Hubei Province
项目面积：120 m²	Project Area: 120 m²
主要材料：金枝玉叶大理石、挪威金大理石、墙布、手绘墙布、马毛皮革、手工地毯、多层黑檀实木地板等	Major Materials: Beige Marble, Norway Gold Marble, Wall Cloth, Hand-painted Wall Cloth, Horsehair Leather, Handmade Carpet, Multi-layered Ebony Solid Wood Floor

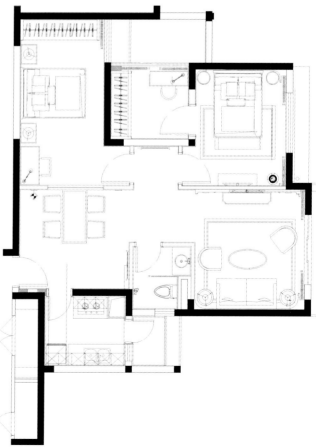

蓝调"Blues"又译为布鲁斯，是一种音乐类型，纯净、舒缓，被称作灵魂音乐。本案设计以"蓝调布鲁斯"为设计灵感。

蓝色给人一种冷静高贵的感受，设计师用浅蓝色碎花图案的盖毯，与居室内的木质元素、深蓝色靠包一起，塑造出一种悠闲自在的家居氛围。居室整体色彩以米、白、灰色等明亮色系作为基础，给人以阳光、开阔的感觉。设计师在浅色调的空间中运用淡淡的蓝色作为点缀，起到了画龙点睛的作用。将清新的蓝色融入到欧式风格的家居装修中，就像源远流长的爵士乐，倾诉着高贵、清澈、有内涵的灵魂故事。

Blues is a kind of music, being pure and relieving, which is called soul music. This project design has blues as the design inspirations.

Blues would produce some cold and noble sensations. The designer makes use of the combination of blankets with light blue floral graphics,

wooden elements inside the space and dark blue package to produce some leisurely home furnishing atmosphere. The whole color of the residence has as base bright colors such as beige, white and gray colors to create some sunny and broad sensations for people. Within the light color tone space, the designer applies light blue as the ornaments, making the finishing point. While integrating fresh blue color into the home furnishing decoration of European style, it is just like the long-standing Jazz music, narrating the noble and clear soul stories of connotations.

凯旋枫丹某住宅
One Residence of Kaixuan Fengdan

设计公司：福建国广一叶建筑装饰设计工程有限公司
方案审定：叶斌
设 计 师：张武
项目地点：福建福州
项目面积：180 m²
摄 影 师：李玲玉
主要材料：意大利大理石、美国木线条、橡木地板、壁纸、PU线条

Design Company: Fujian Guoguangyiye Architectural Decoration and Design Engineering Co., Ltd.
Project Examiner: Ye Bin
Designer: Zhang Wu
Project Location: Fuzhou of Fujian Province
Project Area: 180 m²
Photographer: Li Lingyu
Major Materials: Italian Marble, US Wood Lining, Oak Wood Floor, Wallpaper, PU Lining

MODERN SHOW FLAT III NEW JANE EUROPEAN

本案采用欧式新古典的设计风格，体现成功人士生活的优雅及尊贵。整套空间的规划分为两个部分，动区与静区。外部为客厅、餐厅及玄关区，这是动区。餐厅与厨房呈开放式设计，玄关与餐厅用隐藏式推拉门来划分。客厅在有限的空间中，着重体现一种欧洲情调的乡村感，设计师运用大面积的美国橡木地板来体现这种独特的乡村氛围。

静区主要是卧室区，设计师运用壁纸的纹理及色彩来体现不同年龄段两代人的空间性格，并且合理地运用室内的飘窗来体现卧室的尊贵感。

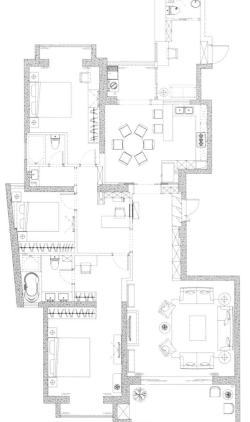

This project applies European Neo-Classical design style to present the elegance and aristocracy of the life of successful people. The planning of the whole space is divided into two parts, the active area and the quiet area. The active area are the living room, dining hall and hallway outside. The dining hall and kitchen have open-style design and the hallway and dining hall is divided from each other with hidden style sliding door. Within the limited space, the living room emphasizes on presenting the country feel of European emotional appeals. The designer makes use of a large area of US oak wood floor to present this peculiar countryside atmosphere.

The quiet area is mainly the bedroom area. The designer applies the texture and colors of wallpaper to display the space features of two generations of people, while appropriately applying the interior bay window to display the noble feel of the bedroom.

依云溪谷169栋

Evian Valley, Building No. 169

设计公司：振勇设计事务所
设 计 师：冯振勇
项目地点：江苏南京
项目面积：480 m²

Design Company: Zhenyong Design Studio
Designer: Feng Zhenyong
Project Location: Nanjing of Jiangsu Province
Project Area: 480 m²

业主是一个对生活有完美要求的人，非常注重细节，对居住空间要求比较高。本案空间的设计有着优雅的线条、柔美的色彩及丰富的层次，使人感受到的不仅是华丽，更是一种不被潮流所淹没的精致感，空间整体上以怀旧风格为主。

设计师将简化的西方古典元素与现代人的审美理念相融合，稳重的对比色调贯穿于整个空间中，局部运用花形图案，衬托出唯美的空间气质。另外，强调精致的细部设计，适度运用光线引导空间动线，搭配具有新古典风格的家具，彰显出具有高贵气质的设计，营造出现代奢华的空间视觉效果。

MODERN SHOW FLAT III NEW JANE EUROPEAN

The property owner is a person who wants life to be perfect, who focuses on details and has high requirements for residential space. The design of this project has elegant lines, soft colors and abundant layers, which make people not only feel the luxury, but also the delicate feel that can not be hidden by tide. As a whole, the space focuses on nostalgic style.

The designer integrates simplified western classical elements with modern people's aesthetic concepts. The sedate contrasting colors go through the whole space, with floral pattern in some parts, setting off the aesthetic space temperament. Other than that, the project stresses on delicate detail design, properly applying rays of light to guide the space moving lines and using Neo-Classical furniture, to set off the design with noble temperament and to produce space visual effects of modern luxury.

君临南山某住宅
One Residence of La France Romanlique

设计公司：江苏镇江乙品环境艺术工程有限公司
设 计 师：方军
项目地点：江苏镇江
项目面积：210 m²
摄 影 师：金啸文

Design Company: Zhenjiang Yipin Environmental Art Engineering Co., Ltd.
Designer: Fang Jun
Project Location: Zhenjiang of Jiangsu Province
Project Area: 210 m²
Photographer: Jin Xiaowen

这个案例户型结构不是很理想，在客、餐厅的中间部位有个卫生间，基本上切断了客厅与餐厅的视线，厨房与餐厅非常拥挤。不过，房子的北面有两个超大的阳台，还有个废弃的公共烟道，通过与业主反复沟通，就把这些空间利用了起来，进行了大刀阔斧的改造。将烟道的部位改成了客卫，厨房与阳台相互连通。于是，一楼有了优雅的门厅、宽敞的餐厅、超大的厨房、南北通透的客餐厅和一个朝南的客房。

餐厅顶面有3根很宽、很厚的横梁，下面原本想摆放一张圆形餐桌，但是因为业主的喜好，最后换成了长方形餐桌，顶面相对应的就设计成了六角形的组合，射灯点缀其中，自然分割了客、餐厅。

客厅作为待客区域，要具备简洁明快的设计效果，是设计的重点区域，相较于其他空间要更加体现居室主人的品位及兴趣爱好。在本案中，主要反映在软装摆件上，体现了居室主人对仿古艺术品的喜爱，同时也反映在对壁纸、石材的偏爱及对各种仿旧工艺的追求上。

The house type structure of this project is not quite ideal, the washroom in the middle area of living room and dining hall basically severs the lines of sight between living room and dining hall, and the kitchen and dining hall appear quite crowded. Other than that, there are two huge balconies in the north of the house, and a deserted public chimney flue. Through repeated negotiations with the property owner, these spaces are applied and moved through drastic regenerations. The area of the flue is changed into a guest washroom, with kitchen and balcony connected with each other. Thus, on the first floor, there are elegant hallway, expansive dining hall, grand kitchen, guest dining hall connecting north and south parts of the house, and a guest room facing south.

The ceiling of the dining hall has three broad and thick beams. In the very beginning, the designer wanted to put a round dining table under the beam. Due to the likes of the property owner, it was changed to be a rectangular dining table. There are the corresponding hexagram grouping on the ceiling. With spotlights decorating the space, dining hall and living room are naturally separated from each other.

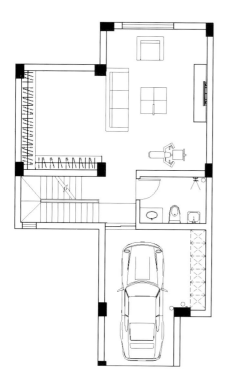
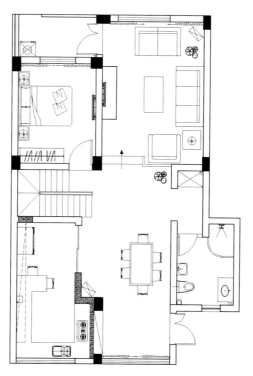
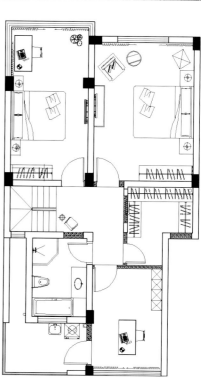

As the guest-receiving area, the living room shall possess the design effects of being concise and brisk. As the design focus, the living room shall further reflect the taste of property owner and his interests compared with other spaces. For this project, the soft decoration objects reflect the property owner's likes for archaized artistic objects, which are also reflected in the preferred wallpaper and stones, together with the various kinds of archaized techniques.

摩登样板间 III
新简欧

钟鼎山庄
Zhongding Villa

设计公司：东易日盛装饰集团
设 计 师：孟繁峰
项目面积：200 ㎡
主要材料：瓷砖、壁纸、布艺、天然石材
摄 影 师：金啸文

Design Company: Dong Yi Ri Sheng Home Decoration Group Co., Ltd.
Designer: Meng Fanfeng
Project Area: 200 m²
Major Materials: Ceramic Tile, Wallpaper, Fabrics, Natural Stone
Photographer: Jin Xiaowen

结构缺陷：本案为200 ㎡的花园洋房，两室朝南，两室朝北，朝北房间面积偏小且采光较差，客厅面积要远大于一般意义上的客厅。餐厅与客厅相比面积较小，且采光差。中厅走廊采光不足。敞开式的北阳台利用率不高，两个卫生间面积差不多，没有功能上的偏重。东门厅兼具过道性质，在空间布局中是难点之一。西门厅厚实的墙面使玄关处的视线受阻，空间比较拘束。另外缺少大户型空间中储存杂物的空间。

风格诠释：根据客户个人的需求定位本案为新古典与后现代的混搭，既有古典的奢华气质，又有现代的明快感。

结构调整：西门厅中，设计师去除了厚实的墙体，替换为双面刻花的镂空屏风，既起到了划分空间的作用，又让视线得以延伸。客厅集会客与休闲为一体，空间层次分明，彼此相融。东门厅采光极好，远眺钟山，景观甚佳，因此设计师在此设置了一处茶饮闲聊的空间，凭窗而望看云卷云舒，一茶一卷品人生百味。针对餐厅与过廊采光较差的问题，设计师将厨房门设计成通透的、装饰性极强的冰雕玻璃门，这

既适当遮挡了厨房，还能将阳光投射进来。更重要的是将北阳台一分为二，一部分成为冷储藏区，家用的冷藏系统和部分杂物有空间放置；另一部分划归在次卫生间，使次卫干湿分离。

色彩设计：设计师以金属感极强的斑驳铁锈色作为背景色，珍珠白勾勒装饰线条，很好地协调了卡其色与白色调的家具。这种搭配让金属与高光的材质得以沉静下来，黑色和烟灰色的配饰点缀其间，使空间色彩丰富却不凌乱，层次分明。主卧室根据女主人自身的喜好，设计师选用了灰紫色作为主基调，色彩典雅，氛围温馨浪漫。

Structure Defects: This project is a garden house of 200 m², with two rooms facing the south and two rooms facing the north, the latter two rooms have small area and bad lighting and the living room's area is far beyond the living room in general meaning. Compared with the living room, the dining hall is small in area and with bad lights. The lighting for central corridor is not sufficient. The use ratio for the open style north balcony is not high and the areas of the two bathrooms are almost the same, with no particular stress on functions. The east hallway has the nature of corridor, as the hard point in space layout. The thick wall of the west hallway blocks the lines of sight of hallway, making the space comparatively restricted. Other than that, the space lacks the storage space for grand house type.

Style Interpretations : This project is oriented to be the mix and match of Neo-Classical and postmodern styles according to the customer's personal requirements, with classical luxury temperament, as well as modern bright feel.

Structural Adjustment: For the west hallway, the designer removes the thick wall, which is replaced with hollow-out screens of double surface carvings, not only dividing the space, but also extending the lines of sight. The living room integrates guest-receiving and leisure, with distinct space layers, which are interconnected with each other. The east hallway has fine lights, when you look towards the Zhong Mountain, you can have perfect views. Based on that, the designer set here a space where people can drink some tea and chat with each other, looking outside the window to observe varying clouds and drinking tea and reading a book to taste the life. For the issues of bad lighting for dining hall and corridor, the designer designed the kitchen door to be transparent and highly decorative ice carving glass door, which not only hides the kitchen, but also invites sunshine inside. What's more important, the north balcony is divided into two, one part is used as refrigerating storage space to include domestic refrigerating system and some sundries, and the other part is used as a small washroom, dividing dry area from the wet area.

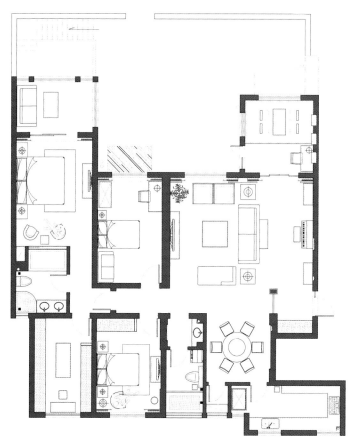

Color Design: The designer uses mottled rust color of great metal feel as the background color and uses pearl white marble to draw the decorative lines, which finely coordinate with the furniture of Khaki and white color tones. This collocations make materials of metal and high luster become quiet, with ornaments of black and smoky gray colors decorating inside, making the space colors become abundant but not messy, with distinct gradations. The master bedroom is designed according to the likes of the hostess herself. The designer selects gray purple color as the tone, displaying elegant colors and warm, romantic atmosphere.

品味生活
To Taste the Life

设计公司：昆明中策装饰（集团）有限公司
设 计 师：窦弋
项目面积：140 m²

Design Company: Kunming Zhongce Decoration (Group) Co., Ltd.
Designer: Dou Yi
Project Area: 140 m²

这是一套140 ㎡的住宅，设计呈现出新古典的空间表情。设计师在满足主人对空间基本需求的情况下展开对空间性格的思索，整体设计采用木纹灰理石作为地面与墙面的主要材料，一体化的材料使空间显得统一而有气势。局部再搭配黑色、白色、褐色、红色、黄色的点缀，将空间沉稳、雅致的性格特征进行了非常到位的表达。这样的搭配使整个空间看起来通透而开阔。再搭配上古典、华丽的家具，稳重而又品位。

This is a residence of 140 m², with design representing the space's Neo-Classical expressions. While meeting with property owner's basic requirements towards the space, the designer carries out thinking towards the space characteristics. The whole design applies wood grain gray marble as the major materials for wall and ground. The integrative materials make the space appear consistent

and magnificent. Accompanied with ornaments of black, white, brown, red and yellow colors in some parts, the sedate and elegant features of the space attain quite appropriate expressions. The collocations makes the whole space appear transparent and spacious. Accompanied with the classical and magnificent furniture, the space appears dignified and graceful.

半岛之恋
Love of Peninsula

设计公司：大墅尚品·由伟壮设计
设 计 师：花苑
软装设计：翁布里亚专业软装机构
软装设计：王一飞
项目面积：350 m²
主要材料：大理石、雪弗板隔断、石膏柱、壁纸等

Design Company: Dashu Shangpin•Zhuang Design
Designer: Hua Yuan
Soft Decoration Designer: Umbria Professional Soft Decoration Institution
Soft Decoration Designer: Wang Yifei
Project Area: 350 m²
Major Materials: Marble, PVC Expansion Sheet Partition, Gypsum Column, Wallpaper

《半岛之恋》这首曲子讲究中国笛与西洋排笛的结合，再辅以钢琴、贝斯为主的爵士乐，笛的音色替代了萨克斯，勾勒出具有东方色彩的旋律主线。

这套方案和这首曲子一样，中西、现代、古典非常和谐地融合到一起，而不拘泥于单一的表现形式。设计上遵循功能至上的原则，色彩以明亮的暖色系为主，使整个空间的氛围更加和谐。设计师摒弃了复杂的造型，纯粹地运用材质并借助灯光将其美感完全地释放出来。空间布局上也稍微做了些调整，将整个空间感表现出来，使其更加符合业主的需要。

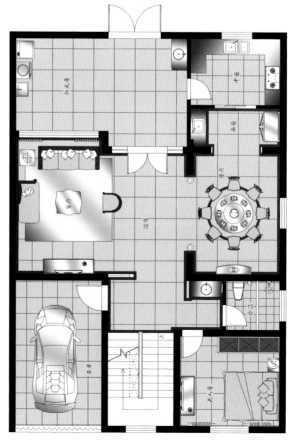

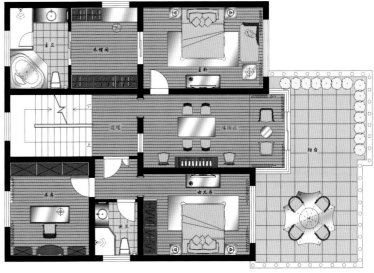

MODERN SHOW FLAT III NEW JANE EUROPEAN

The melody Love of Peninsular strives for the combination of Chinese whistle and western pan flutes, accompanied with jazz music focusing on piano and Bass. The tone of whistle replaces saxophone, creating the leading melodious lines of oriental colors.

This scheme is the same as this melody, while perfectly integrating China and west, modern and classical charms, not being confined to a single expression format. The design follows the principles of supreme functions, while the colors focus on bright warm colors, making the whole space much more harmonious. The designer abandons the complicated format, yet purely applying materials and fully releasing the aesthetic beauty through lights. There are also some tiny adjustments towards the space layout, displaying the whole space feel, which is more in accordance with the requirements of the property owner.

时尚新古典

Fashion New Classical Style

设计公司：北京风尚印象装饰有限责任公司
设 计 师：秦海峰、司洋、吴春艳、赵瑶瑶
软装陈设：尚饰界　SANCY
项目地点：黑龙江哈尔滨
主要材料：多用实木、丝、纱、织物、壁纸、玻璃、仿古瓷砖、大理石等

Design Company: Beijing Fashion Impression Decorations Co., Ltd.
Designers: Qin Haifeng, Si Yang, Wu Chunyan, Zhao Yaoyao
Soft Decoration Furnishings: SANCY
Project Location: Harbin of Heilongjiang Province
Major Materials: Multipurpose Solid Wood, Silk, Yarn, Fabrics, Wallpaper, Glass, Archaized Tile, Marble

本案设计为时尚新古典风格。色彩带动了空间的灵动感，唯美的细节处理、精致的水晶吊灯、华丽的时尚元素都将贵族气息推向了极致，结合传统元素及现代工艺材质、玻璃、花艺、挂画等饰品的巧妙搭配，体现出贵族般的奢华气息。

色彩上，使用黄色、蓝绿色、黑白灰等色调，加上少量黄蓝色的糅合，使整体色彩设计看起来明亮、大方，整个空间给人以开放、宽容的非凡气度。

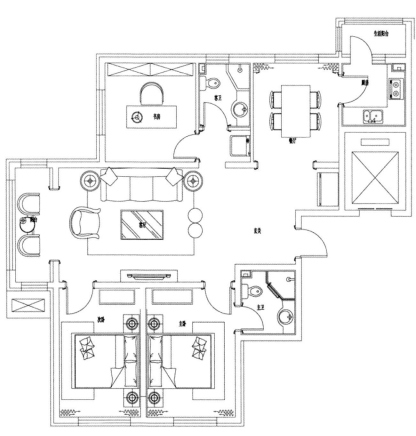

The project design is fashion Neo-Classical style. The colors bring about the dynamic feel of the space. The aesthetic detail treatment, delicate crystal droplights and magnificent fashionable elements promote the noble atmosphere to the extremes. Combined with traditional elements and ingenious collocations of ornaments such as modern crafts materials, glass, floriculture and hanging paintings, the design displays aristocratic luxurious atmosphere.

As for colors, the designer applies color tones such as yellow, aquamarine and black white gray, combined with some yellow blue color, making the whole color design appear bright and grand, while leaving the whole space with open and uncommon bearing, which does not appear narrow at all.

中电颐和家园
Zhongdian Yihe Home

设计公司：北岩设计
项目面积：96 m²
主要材料：彩色乳胶漆、大理石、复合地板等

Design Company: Beiyan Design
Project Area: 96 m²
Major Materials: Colorful Emulsion Paint, Marble, Laminate Flooring

本案业主由于工作需要，经常出差，留宿于各大酒店，各种居室风格都体验过。频繁的出差生活，让他更渴望回家后能安静地读书、写字、听音乐，拥有属于自己的静谧空间。这让我想起《张爱玲传》里一句话："岁月静好，愿使现世安稳。"业主渴望的就是那份恬淡的宁静吧！

房子是三室两厅的格局，面积不大。在结构上需要做出一些调整，北面的飘窗纳入厨房，拆除走道的墙体，增加入户后的空间感，将客厅墙体延伸，保证电视与沙发居中，也增加了客厅空间的舒适度。北面的卧室可以兼作书房及储藏室，这样划分空间比较合理，且简单、适用。

主体空间的色彩选用温和的大地色，客厅舒适的灰绿色布艺沙发，特别选配了复古绿的真皮单人位沙发，给人以内敛的华丽感。深咖色茶几配合墙面上深咖镜框的装饰画，安静而协调。有人会问，为什么餐厅桌椅这般质朴呢？这是我们矛盾的纠结，质朴本色的餐桌椅却选择与华丽的水晶灯相搭配，也是一种矛盾中的和谐感吧。

As job demands, the property owner of this project is always on a business trip and stays in various kinds of grand hotels, and he has experienced various residential styles. The frequent business trip life makes him long more for going back home and enjoying the time reading, writing and listening to music, owning his own serene private space. This reminds me of a famous sentence from writer Zhang Ailing's biography, "Sincerely wish the time be smooth and tidy, and the years be quiet and good." What the property owner longs for is that tranquility indifferent to fame or gain.

The house layout is three bedrooms and two halls, not big in area. The designer needs to make some adjustments towards the structure, the bay window in the

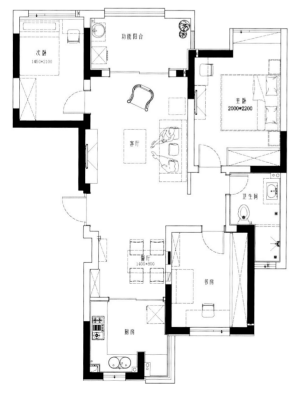

north being incorporated into the kitchen, the corridor wall being removed, space feel at the entrance being strengthened, the living room's wall being extended, and the TV set and sofa kept in the middle area, which add to the comfort feel of the living room space. The bedroom in the north can be study and storage room at the same time, which space separation style being quite reasonable, and simple and practical at the same time.

The color of the main space selects warm ground color and the comfortable gray greenish cloth sofa in the living room specifically selects archaized green leather single sofa, creating some restrained magnificent feel for people. The dark coffee color tea table is accompanied with the decorative painting of dark coffee color frame on the wall, being tranquil and coordinated. Some might ask that why the dining room table and chairs are so primitive. That is the contradictory part we are confused with. The primitive dining table and chairs are allocated with grand crystal lights, and that is the harmonious part in the contradictions.

中南世纪城

Central Living District

设计公司：大墅尚品·由伟壮设计
设 计 师：王璇英
软装设计：翁布里亚专业软装机构
施工单位：大墅施工
项目面积：140 m²
主要材料：地砖、壁纸、石膏线、地板

Design Company: Dashu Shangpin·Zhuang Design
Designer: Wang Xuanying
Soft Decoration Designer: Umbria Professional Soft Decoration Institution
Construction Company: Dashu Construction
Project Area: 140 m²
Major Materials: Floor Tile, Wallpaper, Plaster Lining, Floorboard

简欧是欧式装修风格的一种，多以象牙白为主色调。清新的简欧风格，非常符合中国人内敛的审美观念。本案从简单到繁杂、从整体到局部，精雕细琢，镶花刻金都给人一丝不苟的印象。可以很强烈地感受传统的历史痕迹与浑厚的文化底蕴，同时又摒弃了过于复杂的肌理和装饰，简化了线条。

设计师对每个空间的承接关系都表现得十分到位。黄色调使人们头脑清晰，是稳重的年轻人喜欢的颜色。空间中的摆设搭配更是别出心裁，主要以时尚简约的古典风格饰品为主，将空间装点的不仅有情趣，也营造出浓浓的欧美风情。通过这些简单大方却又可爱浪漫的饰品的装点，也衬托出了主人的爱好与不俗的品位。

Jane European Style is one kind of European decoration style, mainly focusing on ivory white. The fresh Jane European style is quite in accordance with Chinese people's restrained aesthetic standards. This project is scrupulous towards each detail and gives careful revision towards each part, from simple parts to complex parts, from integrity to parts. People can greatly sense the traditional historical traits and the profound cultural connotations, while there are not much too complicated texture and decorations, with lines simplified.

The continuing relationship among spaces is quite proper. Yellow color tone can make people have clear minds, as a color that sedate young people like. The decoration collocations inside the space is quite ingenious, mainly focusing on fashionable and concise objects of classical style. The space is endowed with much more fun, but also creates profound European and American charms. These simple and generous objects yet with lovely and romantic features set off the host's likes and uncommon tastes.

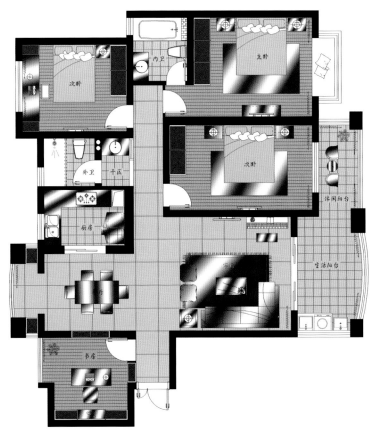

金丰复式 Jinfeng Duplex Mansion

设计公司：汕头市丽景装饰设计有限公司
设 计 师：李伟光
项目面积：210 m²

Design Company: Shantou Lijing Decoration Design Co., Ltd.
Designer: Li Weiguang
Project Area: 210 m²

本案系高级住宅样板间，复式结构。设计师以现代奢华风格来表现空间气质，以此抒发空间主人的追求和品位。

整个空间色调纯粹而饱满，白色、银灰、黑色及咖啡色等偏冷的色彩让空间流露出一种清雅的气质，加之玻璃这种材料的朦胧特性和不锈钢的金属质感，更添几分清冷、明艳。而设计师要表达的就是这样一种概念——"冷暖"，也可以说是一种生活态度，只是藉由空间语言表现出来。优雅舒适的沙发、晶莹剔透的水晶吊灯、熠熠生辉的金属线条、蜿蜒盘旋的楼梯、仿古怀旧的家具、光滑可鉴的地板、华丽庄重的布艺、精妙绝伦的工艺品……冰冷的质感，却表现出一个富有人气的空间，如此"冷暖"，自在人心。

This project belongs to high-level residential show flat, duplex structure. The designer makes use of modern and luxury style to present space temperament, thus displaying space owner's pursuits and tastes.

The color tone of the whole space is pure and profound. The cold colors of white, silver gray, black and coffee make the space send out some elegant temperament. In addition, the hazy features of glass and the metal texture of stainless steel create some chilly, bright and beautiful senses. And that is the concept that the designer wants to present — "cold and warm," which can also said to be some life attitude, being presented with space language. Elegant and cozy sofa, crystal-clear chandeliers, dazzling metal lines, spiral staircase, nostalgic and archaized furniture, smooth floorboard, magnificent and solemn fabrics, delicate artworks... Cold texture presents a space full of popularity. It is all up to you to search out the cold and warm sphere inside the space.

摩登样板间 III
新简欧

时尚圆舞曲 Fashion Waltz

设计公司：摩登时尚装饰设计
设 计 师：刘家洋
项目地点：福建福州
项目面积：300 m²
主要材料：莎安娜大理石、仿古砖、壁纸、实木等
摄 影 师：李玲玉

Design Company: Modern Fashion Decorative Design
Designer: Liu Jiayang
Project Location: Fuzhou of Fujian Province
Project Area: 300 m²
Major Materials: Iran Marble, Archaized Brick, Wallpaper, Solid Wood
Photographer: Li Lingyu

新古典主义风格是由传统的古典主义衍生而来，随着时代的变化，设计师不断为其注入新的设计元素与时代活力。其要义在于以现代的材质与表现手法来赋予空间以古典的韵味，优雅、精致而富有内涵是新古典主义风格的特点。

本案设计充分体现了新古典风格的设计要素。在空间布局上，原本的空间结构合理而完善，设计师在此前提下，以简洁流畅的线条勾勒空间，让空间在平面上得到无限延伸，在空间内部丰富了空间层次，让空间表情更为生动。在材料运用方面，运用了大理石、各种硬包、软包、饰面板、壁纸、实木等，使不同的空间拥有了不同的表情和气质。而这些，最终都围绕着一个空间主题——低调新古典。

MODERN SHOW FLAT III NEW JANE EUROPEAN

Neo-Classical Style derives from traditional classicism. With the change of time, the designer constantly instills some new design elements and time vitality. The key point lies in entrusting the space with classical charms via modern materials and techniques of expression. The features of Neo-Classical Style can be summarized as being elegant, exquisite and full of connotations.

This project design fully displays the design elements of Neo-Classical Style. In space layout, the original space structure is proper and consummate. Under the circumstances, the designer delineates the space with concise and fluent lines, limitlessly expanding the space in plane, enriching the space layers inside the space and making the space expressions more vivid. As for the aspect of materials applications, the designer makes use of marble, various kinds of soft rolls, hard rolls, veneer, wallpaper and solid wood to make different space possess different expressions and temperaments. And finally, all these center on one space theme -- Low-profile Neo-Classicism.

MODERN SHOW FLAT III NEW JANE EUROPEAN

御江金城
Riveria Royale

设计公司：冯振勇室内设计工作室
设 计 师：冯振勇
软装设计：杨钰昕
项目地点：江苏南京

Design Company: Feng Zhenyong Interior Design Firm
Designer: Feng Zhenyong
Soft Decoration Designer: Yang Yuxin
Project Location: Nanjing of Jiangsu Province

本案原为四室两厅两卫户型，改造后成为一个大套间和两个次卧室，常住人口是三位，业主是一对中年夫妻与一个正在读高中的女儿。业主从事媒体工作，经常会出差，所以希望家里以舒适、实用为主，不要过分强调风格，故选择了舒适性、包容性较强的简欧风格。

空间布局上讲究空间的完整性与实用性，选用富有质感的材料、简约的造型及温馨的色调，体现出低调的奢华简约之风。

在室内陈设上，设计师精心挑选了与主人身份、性格爱好相吻合的家具、装饰品，唯美风格的大吊灯与现代简洁的餐桌椅、沙发相呼应，恰到好处地营造出一个宁静、高贵、奢华的极致空间。

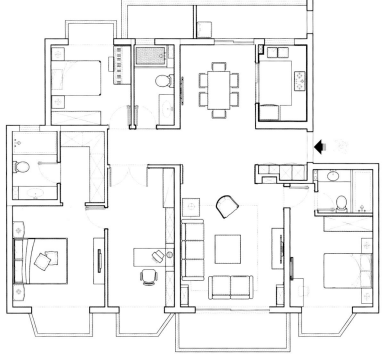

The house type of this project has four bedrooms, two halls and two washrooms. After regeneration, it became a large suite with two little bedrooms. The permanent family number is three, a middle-aged couple and their daughter attending senor hight school. The property owner is engaged in media work and is often on business trip, thus he wishes the home to be comfortable and practical, not to stress on styles. Thus he selects Jane European Style which is comfortable and inclusive.

The space layout stresses on the space integrity and practicality and selects materials of texture, concise format and warm color tone to display low-profile luxury concise style.

As for interior layout, the designer carefully selects furniture and ornaments in accordance with the property owner's status and characteristics. The grand droplights of aesthetic style echo the modern concise dining table and chairs, and the sofa, perfectly creating a tranquil, noble and luxurious space.

四季花园
Four Season Garden

设计公司：巫小伟·威利斯创意设计中心
设 计 师：李琮
软装设计：微诗软装设计
项目面积：89 m²
主要材料：黑檀木钢琴烤漆、软包、壁纸、石膏板、米色大理石、明镜、橡木饰面板等

Design Company: Wu Xiaowei·Willis Innovation Design Center
Designer: Li Cong
Soft Decoration Designer: Wishi Soft Decoration Designer
Project Area: 89 m²
Major Materials: Black Ebony Piano Stoving Varnish, Soft Roll, Wallpaper, Plasterboard, Beige Marble, Bright Mirror, Oak Veneer

本案是公寓房的设计，风格定位为现代简欧。现代主要是给人时尚的感觉，这套设计在时尚的基础上给人带来了华丽、优雅的轻欧风情。

本案原先是两室两厅的格局，业主是一对小夫妻，所以在空间上做了很大的调整，做了一个大套房，内置一个大的卫生间，这样住起来更加舒适。

整体空间以黑色调为主，浅色调穿插其间。黑檀木钢琴烤漆给人华丽、炫目的感觉，明镜的使用使得整个空间更加的通透、敞亮。根据业主需求设计的开敞式的厨房、吧台，让生活更加的充实、惬意。

卧室与客厅在色调上具有极大的反差，橡木色的饰面板本身可以给人温馨的感觉，软包和装饰画使整个空间活灵活现，再通过灯带、壁灯等的点缀，使整个卧室充满温馨、雅致的气息。

This project is design for apartments, with modern Jane European Style. It mainly stresses on creating fashionable sensations for people and this project brings some grand and elegant European charms for people based on fashion. This project used to have the pattern of two bedrooms and two halls. The property owner is a young couple who makes grand adjustments towards the space, such as a grand suite and a large washroom inside, which makes living here more comfortable.

The whole space focuses on black color tone, with light color tones

interlacing inside. Black ebony piano stoving varnish presents people with some glamorous and gorgeous sensations. The bright mirror makes the whole space become more transparent and spacious. The spacious kitchen and bar counter designed according to property owner's requests, life is made more pleasing.

The bedroom and living room have sharp contrast in color tone. The oak color veneer can produce some warm feelings for people. The soft rolls and decorative paintings make the whole space become vivid and dynamic and create some warm and elegant atmosphere for the whole bedroom with lights, wallpaper and other ornaments.

风华年代
The Time of Elegance

设计公司：一空设计事务所
设 计 师：沈一
项目地点：浙江杭州
项目面积：150 m²
主要材料：热带雨林大理石、橡木地板、仿古砖、进口壁纸

Design Company: YKON Design Studio

Designer: Shen Yi

Project Location: Hangzhou of Zhejiang Province

Project Area: 150 m²

Major Materials: Tropical Forest Marble, Oak Floorboard, Archaized Brick, Imported Wallpaper

本案是简欧风格的案例,从奢华到简洁、从整体到局部,精雕细琢、雕花刻金都给人一丝不苟的印象。局部用现代的表现手法,光影、材质与色彩的搭配,赋予空间以层次感与节奏感,人性化的功能分区与整体空间完美搭配,营造出低调奢华、沉稳华贵的空间气质。

设计者认为艺术来源于生活,反过来又作用于生活,只有经过生活洗礼的艺术才是真正为大家所接受的,也才能经得起时间的考验。

This project is Jane European Style and leaves people with meticulous impressions from luxury to conciseness, from whole to parts, with delicate drawing and careful revision. The detail parts apply modern expression techniques to entrust the space with layer and rhythmic feel, through collocations of light and shadow, materials and colors. The human functional division is in perfect match with the whole space, creating some sedate and noble space temperament of low-profile luxury.

The designer believes that art is rooted in life and acts on life. Only the art through baptism of life would be truly accepted by all, and would stand the test of time.

富丽湾何公馆
Fuli Bay He's Mansion

设计公司：宁波江北 UI 室内设计有限公司
设 计 师：陈显贵
项目地点：浙江宁波
项目面积：300 m^2
摄 影 师：牛斐

Design Company: Ningbo Jiangbei UI Interior Design Co., Ltd.
Designer: Chen Xiangui
Project Location: Ningbo of Zhejiang Province
Project Area: 300 m^2
Photographer: Niu Fei

花香静苑

这是一套兼具冷静与优雅品质的私家住宅，与女主人的知性气质相得益彰，本案的女业主有非常好的品位与审美，这让设计师与她有很好的共鸣。

当铜艺花灯的光辉悄然洒落，鲜花绽放于桌几，客厅萦绕着浪漫与优雅的气息。墙面与天花的设计素雅且温馨，但精致的细节却一丝不苟。公共空间仿古地板的颜色恰到好处，仿佛历经岁月洗礼后依然尊贵而美丽，凸显出居室的文化气息与主人的沉静气度。楼梯下，一幅画、一张椅，就那么静静地倚在门边，似岁月留痕，光华荏苒。空间中没有刻意装饰的痕迹，无论是装饰画、窗帘、绿植还是摆件与陈设，都显得十分的自然、协调。

在卧室中舒适的窗前小座，可以让人睡前小叙，也可以享受清晨阳光，放松身心，边桌上有仿古电话机和老旧的铜灯，仿佛让人置身于昔日的上海，绽放花样年华。

Fragrant and Tranquil Mansion

This is a private mansion integrating calmness and elegant quality, with the hostess' intellectual temperament bringing out the best in each other. The hostess of this project has great taste and aesthetic capability, which makes the designer have fine resonance with her.

When the splendor of copper art festive lantern scattered around, the flowers are blossoming on the table and the living room would be filled with romantic and elegant atmosphere. The design of wall and ceiling is elegant and warm, but the exquisite details are just meticulous. The color of antique floorboard inside the public space is to the point, seeming to remain the nobility and elegance after so many years and highlighting the residence's cultural ambiance and the owner's sedate bearing. Under the stairs, a painting and a chair are set beside the door quietly, just like something left by the time and representing the elapsing time. Inside the space, you can not find the

trait of deliberate decoration, and the paintings, curtains, green plants, the decorative objects and furnishings all appear so natural and coordinated.

Taking a seat before the window of the bedroom, having some chat before sleep, or enjoying the morning sunshine, you can get relaxed physically and mentally. On the side table are archaized telephone and antique copper light, all make people feel like being in Shanghai of old days, and enjoy the blossoming age.

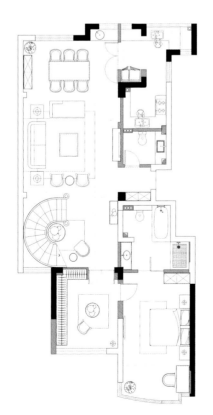
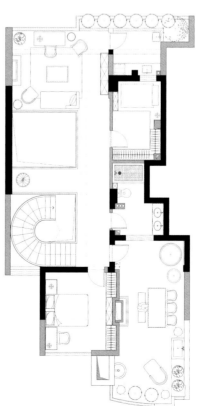

MODERN SHOW FLAT III NEW JANE EUROPEAN

怡湖新贵 Yihu New Aristocracy

设计公司：北岩设计
项目地点：江苏南京
项目面积：340 m²
摄　　影：金啸文空间摄影

Design Company: Beiyan Design
Project Location: Nanjing of Jiangsu Province
Project Area: 340 m²
Photographer: Jin Xiaowen Space Photography

新贵主义风格是当代以现代欧式主义风格为基调，并融入其他风格元素的空间形态，具有强烈的时代特征。它以抽象的古典元素为基调，用现代材料完成几何轮廓的构造，以简洁的线条装饰、明亮的色彩对比为装饰特征。

避免有简单形式，无文化元素

壁炉是欧式风格设计的主要元素之一，相对应的椅子、壁灯等软装饰品，是不可或缺的元素，这几个元素的造型、材质、色彩、体量感等都是整体空间风格走向的风向标。当然随着时间的推移，壁炉已经成为饰品，而相应的碳箱、碳钳、碳钎等物品也渐渐消逝，但壁炉的空间"扮靓"功效仍会存在下去。

在该案中，设计师用现代的手法和材质还原古典气质，整个空间的设计具备了古典和现代的双重审美效果，这种效果与现代人所追求的品质与休闲生活的完美结合相适应，使业主在享受物质生活的同时，也得到了精神上的慰藉。

New Aristocracy Style is a space format in contemporary world with modern European style as the tone and integrating the elements of other styles, with profound time features. Its keynote is the abstract classical elements and its decorative features are the construction of geometric outline with geometric outline and concise line decorations, together with the sharp contrast of bright colors.

Avoiding Simple Forms, and No Cultural Elements

Furnace is one of the main elements of European style design and the corresponding chairs, wall lamps and other soft decorative objects are the indispensable elements. These elements' format, materials, colors and scales are the indicator of the whole space's style orientation. With the moving of time, these wall lamps have become ornaments and the relevant carbon box, carbon pliers and other objects are disappearing gradually, but the space "beautifying" function of furnace would remain for long.

For this project, the designer makes use of modern approaches and materials to restore classical temperament and the design of the whole space possesses the dual aesthetic effects of classical and modern aspects, which is quite in accordance with the perfect combination of taste and leisure life that modern people pursue, making people get spiritual comfort while enjoying the material life.

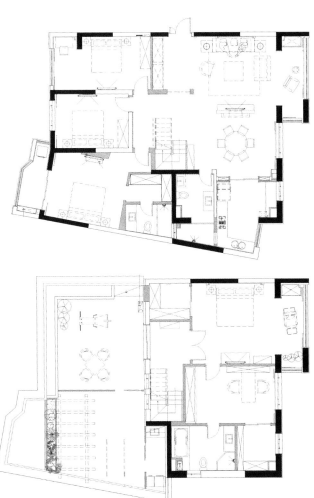

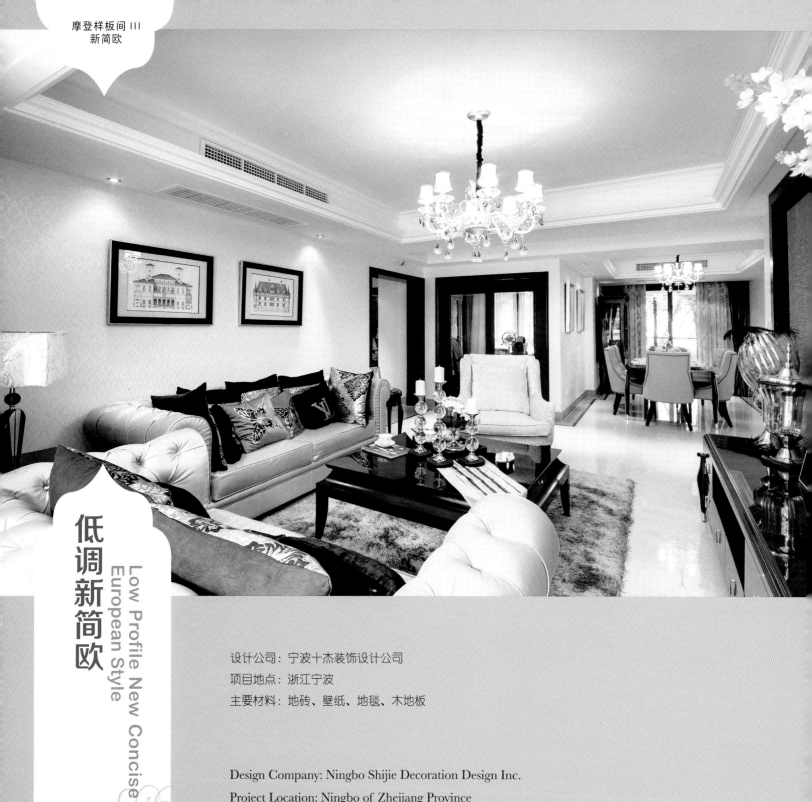

低调新简欧
Low Profile New Concise European Style

设计公司：宁波十杰装饰设计公司
项目地点：浙江宁波
主要材料：地砖、壁纸、地毯、木地板

Design Company: Ningbo Shijie Decoration Design Inc.
Project Location: Ningbo of Zhejiang Province
Major Materials: Floor Tile, Wallpaper, Carpet, Wood Floorboard

MODERN SHOW FLAT III NEW JANE EUROPEAN

本案想传递一种低调奢华的生活方式，设计师将欧式生活的贵气与精致带入生活中来，却又无金碧辉煌的浮华感，无论坐在客厅还是花园小憩，一杯浓浓的咖啡、一本小说，都是一种意境，一种浪漫的情怀……

整套大宅的设计是欧式奢华主义与现代浪漫主义的完美结合，设计师采用流线型与直线相互搭配，更显家具的现代奢华感。明亮的色彩在室内相互交错，豪华的家饰与充满个性的壁纸相互呼应，此情此情充斥着浪漫的诱惑力与华贵感，表现出大气的王者风范。

This project wants to present some low-profile and luxurious life style. The designer integrates inside life the noble and delicate charms of European life, but with no resplendent or ostentatious feel. When you sit inside the living room or take a rest inside the garden, you can enjoy a cup of strong coffee or read a novel, there is the artistic conception, and some romantic emotions...

The design of the whole mansion is the perfect combination of European luxury and modern romanticism. The designer applies the mutual combination of streamline and straight lines, further displaying the modern luxury feel of furniture. Bright colors are interlacing with each other inside the space and the luxurious furnishings and peculiar wallpaper echo each other, which is full of romantic attraction and noble feel, displaying some magnificent atmosphere.

江景大厦某住宅

One Residence of Riverview Building

设计公司：海璟道设计顾问（香港）有限公司
设 计 师：刘海波
项目地点：湖北武汉
项目面积：400 m²
主要材料：石材、护墙板、壁纸、地毯
摄 影 师：罗施

Design Company: Haijingdao Design Consultants (Hong Kong) Co., Ltd.
Designer: Liu Haibo
Project Location: Wuhan of Hubei Province
Project Area: 400 m²
Major Materials: Stone, Wainscot Board, Wallpaper, Carpet
Photographer: Luo Shi

本案中，设计师以传统的庄重典雅作为空间的整体设计思想，使室内空间具有更多的舒适性和实用性。入门处的玄关，璀璨的紫水晶、线绒的布艺沙发、配以色彩清丽而精致典雅的摆件和挂画，共同塑造出了一个令人宾至如归的客厅空间。

居住空间是体现居室主人性格、爱好的地方，在设计上就要表达出来。因此，在本案中，设计师将重点放在软装饰的布置上，墙面整体铺贴花朵图案的壁纸，顷刻间就营造出温馨、典雅的主基调。枣红色木质餐桌椅、墙裙、书柜等，又塑造出空间的尊贵感。

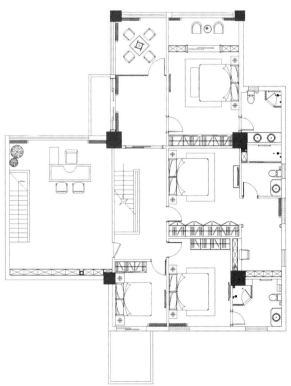
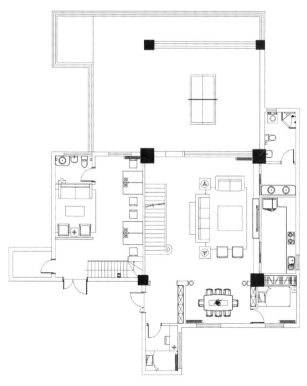

For this project, the designer uses traditional sedate elegance as the whole design concept of the whole space, making the interior space become more comfortable and practical. For the hallway at the entrance, the resplendent amethyst, fabric cloth sofa and elegant ornaments and paintings of bright colors together create a pleasing living room space.

The residential space is representing the property owner's characteristics and likes, which shall be represented in design. Thus, for this project, the designer focuses on the setting of soft decorations and the wall is decorated with wallpaper of floral pattern, creating warm and graceful tone immediately. Purplish red wooden dining tables and chairs, wainscot and bookcase create some noble feel for the space.

橘郡阳光 Orange County Sunshine

设计公司：鸿扬集团 陈志斌设计事务所
设 计 师：陈志斌
项目地点：湖南长沙
项目面积：280 m²
主要材料：樱桃木墙挂板、仿古砖、皮质软包、米色墙漆、樱桃木地板

Design Company: Hongyang Group, Chen Zhibin Design Firm
Designer: Chen Zhibin
Project Location: Changsha of Hunan Province
Project Area: 280 m²
Major Materials: Cherry Wood Hanging Board, Archaized Brick, Leather Soft Roll, Beige Wall Paint, Cherry Wood Floorboard

橘郡，是一处经典的美式风格楼盘，楼盘借助山地的特殊地形形成了错落的景观，为广大业主带来了丰富的视觉享受。

本案所处位置优越，风景极佳。共有4层，空间布局非常开阔。客厅与餐厅构成了一个大的、开放式的公共空间，楼梯间的开窗与客厅互动，并设置了娱乐室、储藏间、保姆间、酒窖等重要功能区。二、三层为静区，是主人与孩子休息、学习的空间，并有露台以供观景。

在材质的选择上，设计师选用仿古砖、樱桃木、米色墙漆来体现美式风格的经典色调，并辅以柔和的皮质材料，以起到静音的效果。仿古砖的色系深中带浅，棕中配绿，其生动、活泼的空间表情，体现出主人不俗的生活情趣。

Orange County is a property of classical American style, which produces landscapes strewn at random relying on the peculiar mountainous terrain, providing the property owners with great visual enjoyments.

This project enjoys advantageous location and fine views. With four floors, the space layout is quite expansive. The living room and the dining hall form a large and open style public space, the windows of the staircase interact with the living room, while there are important functional areas such as entertainment room, storage room, nanny room and cellar, etc. The second and third floors are the quiet zone, as the rest and learning space for the masters and the children, while there is a balcony for people to enjoy the views.

As for materials selection, the designer selects archaized brick, cherry wood and beige wall paint to represent the classical tones of American style and attains the mute effect, accompanied with soft leather materials. The color of archaized tiles is dark yet with some light sense, green accompanying brown, with its vivid and lively space expressions representing the uncommon life interests of the host.

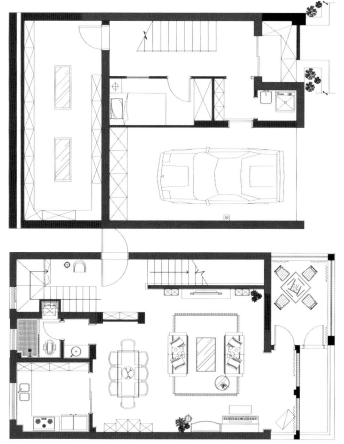

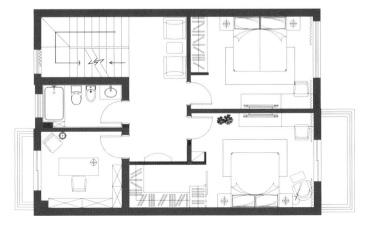

宾王广场样板间

设计公司：文献装饰设计有限公司
设 计 师：吴献文
项目地点：浙江义乌
项目面积：160 m²
主要材料：乳胶漆、大理石、抛釉砖、实木板、壁纸、清玻
摄 影 师：吴献文

Design Company: Wenxian Decorative Design Co., Ltd.
Designer: Wu Xianwen
Project Location: Yiwu of Zhejiang Province
Project Area: 160 m²
Major Materials: Emulsion Paint, Marble, Glazed Ceramic Tile, Solid Wood Floor, Wallpaper, Clear Glass
Photographer: Wu Xianwen

静谧的午后，阳光温暖的问候抵达我的心底。品尝着泡沫咖啡的味道，安静中倾听自己的心跳。凝望远方洁白的云朵，憧憬着彼岸的幸福，幻想那个邂逅的街角，繁花若锦，正如你那灿烂的笑容。在某个相遇的年华里，只愿得到一个白首不分离的心。

本案定位为现代简欧风格装饰设计，主体颜色以白色、咖啡色为主，结合大理石、实木板、抛釉砖等高档装饰材料，并配以壁纸、清玻等装饰元素，材料选择上的精心搭配，就使得空间有了生命力。一进门后，舒缓的感觉就会使你感到无比的惬意。设计师巧妙地把复杂的线条进行抽象简化，用一些新的装饰元素与摆设来体现居室的个性特征，也体现出主人生活的不俗品质和丰富多彩的生活。

In the tranquil afternoon, the warm greetings of sunshine arrive at the bottom of my heart. I taste the foam coffee and listen to my heartbeat quietly. I am staring at the distant white clouds, long for the other bank's happiness and think about that street corner where we met by chance. The blossoming flowers are just like your brilliant smile. At the time when we met, I only hoped that I could own the people who would never leave me.

This project is oriented to be decorative design with modern Jane European Style, with tone colors of white and coffee, combined with high-end decorating materials such as marble, solid wood floor, glazed ceramic tile, etc., together with elements such as wallpaper and clear glass. The ingenious collocations on materials selection allow the space to have vitality. Upon entering the space, the soothing feel would make you feel extremely pleased. The designer ingeniously makes the complicated lines abstract and simplified, representing the residential features with some new decorative elements and ornaments, thus displaying the property owner's uncommon quality and colorful life.

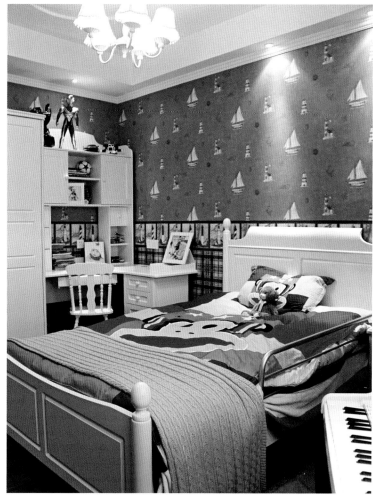

来宾海德堡样板间
Laibin Heidelberg Show Flat

设 计 公 司：深圳太合南方建筑室内设计事务所
设 计 师：王五平
项 目 地 点：广西来宾
项 目 面 积：160 m²
主 要 材 料：乳胶漆、黑镜钢、水曲柳油白、全抛釉砖、壁纸等

Design Company: Shenzhen Taihe South Architectural Interior Design Firm
Designer: Wang Wuping
Project Location: Laibin of Guangxi Province
Project Area: 160 m²
Major Materials: Emulsion Paint, Black Mirror Steel, Ash-tree Oil White Coating, Glazed Ceramic Tile, Wallpaper

本项目是样板示范单位，在平面改动上比较大，总的设计原则是保持空间的大气、通透、舒适。原平面布局中，进门左手边是一个小的入户花园，现在改造成了鞋帽休闲区，在客厅的分界处巧妙地设计了一个休闲吧台加以分割，这样就很好地把客厅和鞋帽休闲区连成了一个整体，突显出一个大面积的客厅空间，吧台还可以作为客厅的沙发靠背。

原平面布局中的餐厅面积不大，所以我们想把它设计成一个餐厨一体的空间，定位于简欧西式的生活理念，既大气又方便。

湖与墅别墅样板间
Villa Life by the Lake Show Flat

设计公司：北京风尚印象装饰有限责任公司
空间设计：胡强
软装陈设：尚饰界 SANCY
软装设计：郝亚筱、司洋、吴春艳
设计顾问：古翔飞、赵丹
项目地点：黑龙江哈尔滨
硬装材料：西班牙米黄（老矿）石材、西班牙米白（新矿）石材、深啡网石材、浅啡网石材、樱桃木做旧、混油墙板、有色墙漆、进口羊毛定制地毯
软装产品：家具、装饰画、水晶吊灯、饰品、花器

Design Company: Beijing Fashion Impression Decorations Co., Ltd.
Space Designer: Hu Qiang
Soft Decoration Furnishings: SANCY
Soft Decoration Designers: Hao Yaxiao, Si Yang, Wu Chunyan
Design Consultants: Gu Xiangfei, Zhao Dan
Project Location: Harbin of Heilongjiang Province
Hard Decoration Materials: Spanish Beige Stone, Spanish Off-white Stone, Emperador Dark, Emperador Light, Antique Cherry Wood, Wallboard, Colored Wall Paint, Imported Wool Custom-made Carpet
Soft Decoration Materials: Furniture, Decorative Painting, Crystal Droplights, Ornaments, Flower Vase

根据楼盘的定位，客户群定位在 30 岁到 50 岁之间的成功人士。他们或有海外的生活感受或经历，或向往轻松而休闲的理想家居生活，同时还期待拥有一个能体现主人身份地位及文化品位的生活空间。

客厅：带有丰富造型和线条层次的壁炉与视听柜既满足了功能的需求也起到了装饰墙面的作用。

书房：书房采用造型细腻、层次变化丰富的白色线条来诠释空间的内涵，顶面的几何分割与墙面古典分割造型，使得整个空间浑然一体。

餐厅：古典的分割形式搭配丰富的装饰线条是本案重要的表现手法。

卧室：丰富的造型及变化统一在同一色系内，干净的底色搭配色彩变化丰富的家具、饰品与布艺，实现了软、硬装两者之间的对立统一。

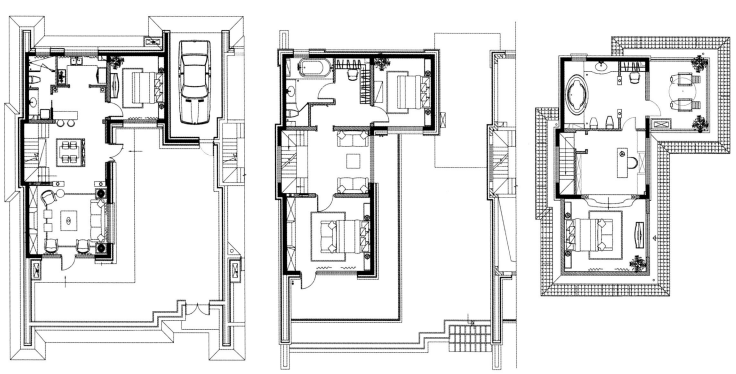

MODERN SHOW FLAT III NEW JANE EUROPEAN **283**

Based on the orientation of the property, the customer group is oriented to be successful people between 30 and 50-year-old. They may have overseas life feelings or experiences, or long for easy and leisurely ideal furnishing life. While at the same time, they expect to have a living space which can represent the master's status and cultural tastes.

Living Room: The furnace and audio-visual cabinet with abundant formats and layers can not only meet with functional requirements, but also play the role of decorating wall.

Study: The study applies white lines of refined formats and rich layer variations to interpret the space's connotations. The ceiling's geometric segmentations and wall's classical segmentation format make the whole space appear like a whole.

Dining Hall: The collocation of classical segmentation form and rich decorative lines is an important presentation of this project.

Bedroom: Rich formats and variations are combined in the same color system, and the clear ground color matches the furniture, ornaments and fabrics of rich color variations, attaining the contrasting integration of soft and hard decorations.

摩登样板间 III 新简欧

中南世纪城9幢
Central Living District Building No. 9

设计公司：大墅尚品·由伟壮设计
设 计 师：由伟壮、王伟
软装设计：翁布里亚专业软装机构
施工单位：大墅施工
项目地点：江苏常熟
项目面积：150 m²
主要材料：仿大理石砖、壁纸、实木地板、白色木质护墙板、油漆等

Design Company: Dashu Shangpin·Zhuang Design

Designers: You Weizhuang, Wang Wei

Soft Decoration Designer: Umbria Professional Soft Decoration Institution

Construction Company: Dashu Construction

Project Location: Changshu of Jiangsu Province

Project Area: 150 m²

Major Materials: Simulated Marble Brick, Wallpaper, Solid Wood Floorboard, White Wood Wainscot Board, Oil Paint

现代欧式风格通常采用深色的橡木或枫木家具和软装饰来营造整体效果。本案中的布艺沙发组合有着丝绒的质感及流畅的木质曲线，将传统欧式家居的奢华与现代家居的实用性完美地结合起来。还有精美的油画，制作精良的雕塑工艺品，都是打造现代欧式风格不可缺少的元素。

本案有的不只是豪华、大气，更多的是惬意和浪漫。通过完美的曲线，精益求精的细节处理，带给人们一种属于家的温暖感觉，传统欧式木质家具色泽庄重，纹理精细美观，其整个框架显露出一番别样的古典韵味。

Modern European style generally applies dark oak or maple furniture and soft decorations to produce the integral effects. For this project, the cloths sofa has velvet texture and smooth wooden curves, perfectly integrating the luxury of traditional European furnishing with the practical modern furnishing. Other than that, exquisite paintings and excellent sculpture artworks are all indispensable elements in creating modern Europeans style.

What this project possesses are not only luxury and magnificence, but also pleasure and romance. Through perfect curves and excellent details treatment, this project creates some warm feelings of home for people. The traditional European wooden furniture has sedate colors and exquisite patterns, which make the whole frame display some peculiar classical charms.

天正桃园
Tianzheng Taoyuan

设计公司：冯振勇室内设计工作室
设 计 师：冯振勇
项目面积：249 m²

Design Company: Feng Zhenyong Interior Design Firm
Designer: Feng Zhenyong
Project Area: 249 m²

该案设计以现代简欧风格为主，色彩冷静，线条简洁而优美。客厅中的金属色与线条感充满了动感的旋律，简洁而不失时尚气息。电视墙的处理简洁而富有品位，金色并带有深浅韵纹的点缀，凸显出空间的尊贵、典雅感。客厅中的深咖色调使居室充满休闲的氛围，是快节奏都市生活的休憩港湾。

也许是觉得这种过于冷静的家居格调显得不够柔和，设计师特意选择了一些温馨的家居饰品来协调、中和这种冷静感。鲜艳的花朵、富有生机的绿植、优雅的配饰，不花哨，也不影响整个居室的平静感，有着画龙点睛的作用。柔和、偏暖色的灯光也让整体素雅的居室不会有太多的冰冷感觉。

餐厅中白色的座椅与深色的客厅形成鲜明的对比，层次分明。餐桌上精致的瓷器、陶艺、色彩和造型丰富多样，增加了空间的高贵感。吊顶的处理也极尽心力，正圆的造型，加上亮面的顶面材质，不显得繁琐，但却很特别。

The project design focuses on modern Jane European Style, with cold colors, concise lines and beautiful appearance. Inside the living room, the metal color and lines are full of dynamic rhythms, being concise and fashionable. The TV background wall is concise and full of taste. The gold connotations with dark and light stripes highlight the noble and elegant feel of the space. The dark coffee color of the living room makes the interior space full of leisurely atmosphere, creating a leisure harbor for the fast-pace urban life.

Maybe the designer thinks that the much too cold furnishing tone does not possess softness, thus he specifically selects some warm furnishing objects to coordinate and neutralize this cool feel. The fresh flowers, dynamic green plants and elegant ornaments are not gaudy at all and do not affect the tranquil feel of the

whole residence, yet making the finishing point. The soft and warm lights make the elegant space do not appear that cold.

Inside the dining hall, the white chairs produce sharp contrast with the dark living room, with distinct gradations. On the dining table, the exquisite ceramic tile and pottery have abundant colors and format, increasing the noble feel of the whole space. The designer makes great efforts towards the treatment of the ceiling. The circular format and ceiling materials of bright surface do not appear much complicated, but with some different feel.

MODERN SHOW FLAT III NEW JANE EUROPEAN

嘉宝田花园
Jiabaotian Garden

设计公司：深圳市伊派室内设计有限公司
项目地点：广东深圳
项目面积：220 m²
主要材料：石材、玻璃、水晶、木材、壁纸

Design Company: Shenzhen Yipai Decoration Co., Ltd.
Project Location: Shenzhen of Guangdong Province
Project Area: 220 m²
Major Materials: Stone, Glass, Crystal, Wood, Wallpaper

豪宅不只是对奢华生活的一种渴望，从居住的本质上讲，人们更希望从中得到生活环境和品质的改善。

本案的室内装饰设计充满着古典、高雅的气息，设计师倡导温馨、浪漫、舒适的生活方式。以新古典为基本的表现框架，融入新的思维理念来连贯客、餐厅及其他空间，搭配运用线板、壁板、斗框等新古典语汇，衬托优雅高挑的空间品质，也使得客、餐厅相互呼应，使空间更具张力。

不难看出设计者尤为注重空间功能的完善，注重发挥结构本身的形式美。造型大气，使用合理的造型工艺，尊重材料的性能，讲究材料自身的质地和色彩的配置效果，寻找洁净的、直截了当的美。

Mansion is not only a desire for luxury life. From the aspect of the nature of residence, people would rather attain the improvement of living environment and quality from it.

The interior decoration design of this project is full of classical and elegant atmosphere. The designer advocates warm, romantic and cozy lifestyle. With Neo-Classical style as the basic expression framework, this project integrated new thinking ideas to connect living room, dining hall and other spaces. The combination of Neo-Classical elements such as filament plates, wallboard, frame, etc., to set off the elegant and high space quality, making living room and dining hall correspond with each other, while endowing the space with more tension.

It is not difficult to find out that the designer especially focuses on perfection of space functions and emphasizes on displaying the structural beauty. The format is quite magnificent, the designer applies proper format techniques, respects the performance of materials and aspires for the materials' quality and colors' collocation effects, searching for clear and straight-way beauty.

梅溪湖 D6 区样板间
Meixi Lake D6 District, Show Flat

设计公司：北京东易日盛装饰
设 计 师：王帅
项目面积：180 m²

Design Company: Dong Yi Ri Sheng Home Decoration Group Co., Ltd.
Designer: Wang Shuai
Project Area: 180 m²

新古典简欧风格不是纯粹旧元素的堆砌，而是通过对传统文化的认知，将现代与传统结合在一起，以现代人的审美要求来打造富有传统韵味的事物，使传统艺术的脉络传承下去。

"形神聚散"是本套样板间的主要特点，设计师采用简化的手法、现代的材料和加工技术去追求传统式样的大致轮廓特点，在注重装饰效果的同时，还原了古典气质，使之具备古典与现代的双重审美标准。

在色调上，优雅的白色是整个空间的主基调，黑色、淡淡的浅褐色、金色贯穿于整个空间中，使整个空间看起来通透且开阔。

Neo-Classical Jane European Style is not the piling-up of old elements, but integrates modern and traditional spheres through acknowledgments of traditional culture, and creates things of traditional charms with modern people's aesthetic requirements, passing on the traditional art.

The designer applies simplified approaches, modern materials and processing techniques to seek for the approximate layout features of traditional style, restoring classical temperament while focusing on decorative effects, and making the space acquire dual aesthetic standards of classical and modern aspects.

As for color tone, elegant white color is the keynote of the whole space, black and light brown color and gold color penetrate through the whole space, making the whole space appear transparent and expansive.

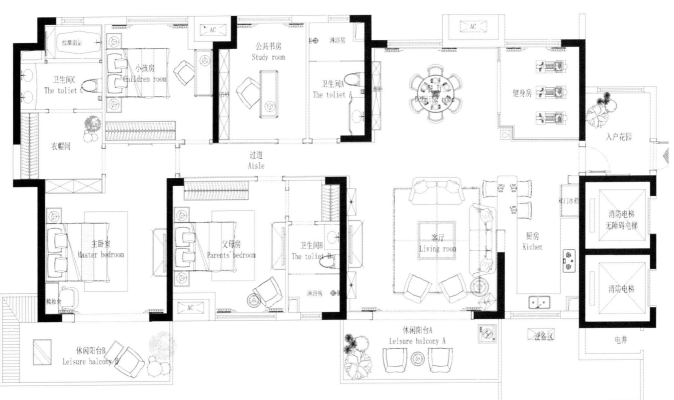

MODERN SHOW FLAT III NEW JANE EUROPEAN

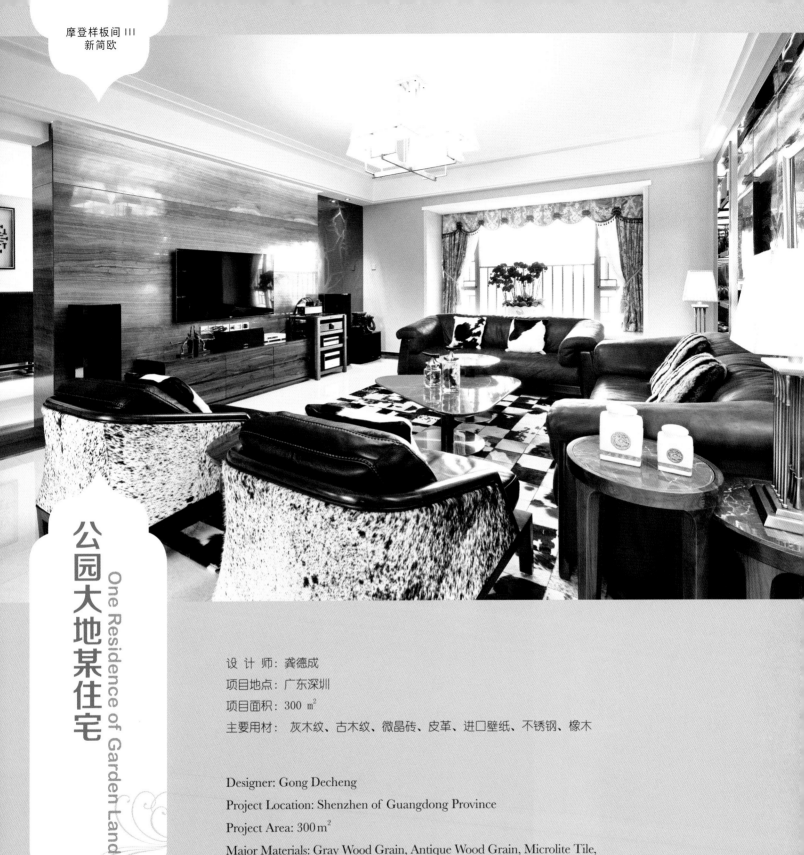

公园大地某住宅
One Residence of Garden Land

设 计 师：龚德成
项目地点：广东深圳
项目面积：300 m²
主要用材：灰木纹、古木纹、微晶砖、皮革、进口壁纸、不锈钢、橡木

Designer: Gong Decheng

Project Location: Shenzhen of Guangdong Province

Project Area: 300 m²

Major Materials: Gray Wood Grain, Antique Wood Grain, Microlite Tile, Leather, Imported Wallpaper, Stainless Steel, Oak

在本套方案的设计中，设计师用简约、现代的设计手法，营造出一个时尚、有品位、具有都市感的现代家居空间。

整体色调以浅色为主，浅色的木饰面、浅灰的壁纸、浅色的地砖，整个色调让人感觉平稳而干净。局部的深色大理石及各具特色的工艺品，在整个浅色背景的衬托下，成为视觉的中心。

材质的选择上我们选用造型简洁而又不失质感的材质，让整个设计看上去简约而不简单。整个设计过程我们都遵循这一准则，为最终的设计效果奠定基础。最终，展现在我们面前的就是一个极具当代都市情调的现代简欧之家。

For space design of this project, the designer makes use of concise and modern design approaches to create a modern furnishing space of fashion, taste and metropolitan feel. The whole color tone focuses on light color, with light color wood veneer, light gray wallpaper and light color floor tiles, which make all appear clear and tidy. Set off by the whole light color background,

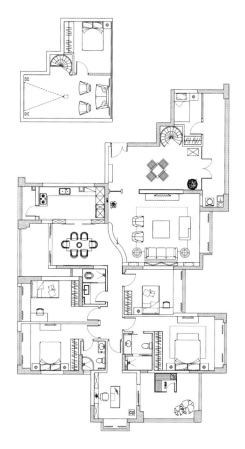

the dark color marble in some parts and the handiworks of various features become the visual focus.

As for materials selections, we select materials of concise format and of fine texture, which make the whole design appear concise but not simple. During the whole design process, we always follow this principle, thus establishing a foundation for the final design effects. Finally, what displays before users is a modern Jane European style home.

本书在编写过程中，得到各位参编老师的倾力协助，特表示感谢，以下为参编人员名单（排名不分先后）：

林冠成	方　峻	叶景星	黄书桓	欧阳毅	王基守	陈佳琪	蔡明宪	胡春惠	胡春梅	邱春瑞
廖易风	孟繁峰	冯易近	陈显贵	刘　鹰	李　跃	黄　杰	张晓莹	马利华	肖飞生	徐树仁
施　凯	黄耀国	刘海涛	庄锦星	叶　斌	温旭武	吴文粒	陆伟英	张　武	李玲玉	冯振勇
方　军	金啸文	窦　弋	王一飞	秦海峰	司　洋	吴春艳	赵瑶瑶	王璇英		

图书在版编目(CIP)数据

摩登样板间. 第3辑. 新简欧 / ID Book 图书工作室编 —武汉：华中科技大学出版社，2014.9
ISBN 978-7-5680-0067-3

Ⅰ．①摩… Ⅱ．①I… Ⅲ．①住宅－室内装饰设计－图集 Ⅳ．①TU241-64

中国版本图书馆CIP数据核字(2014)第100150号

摩登样板间Ⅲ 新简欧　　　　　　　　　　　　　　　　　　　　　　　　　　　ID Book 图书工作室 编

出版发行：华中科技大学出版社（中国·武汉）	
地　　址：武汉市武昌珞喻路1037号（邮编：430074）	
出 版 人：阮海洪	

责任编辑：赵爱华		责任监印：秦　英
责任校对：曾　晟		装帧设计：张　艳

印　　刷：北京利丰雅高长城印刷有限公司
开　　本：965 mm×1270 mm　1/16
印　　张：20
字　　数：256千字
版　　次：2014年9月第1版第1次印刷
定　　价：338.00元　（USD 69.99）

投稿热线：(010)64155588-8000
本书若有印装质量问题，请向出版社营销中心调换
全国免费服务热线：400-6679-118 竭诚为您服务
版权所有　侵权必究